900430152 4

WITHDRAWN
FROM
UNIVERSITY OF PLYMOUTH
LIBRARY SERVICES
KV-495-880

GROUNDWATER & SOIL CLEANUP

IMPROVING MANAGEMENT OF PERSISTENT CONTAMINANTS

Committee on Technologies for Cleanup of Subsurface Contaminants in the DOE Weapons Complex

Board on Radioactive Waste Management
Commission on Geosciences, Environment, and Resources

National Research Council

NATIONAL ACADEMY PRESS
Washington, D.C.

NATIONAL ACADEMY PRESS • 2101 Constitution Avenue, NW • Washington, DC 20418

NOTICE: The project that is the subject of this report was approved by the Governing Board of the National Research Council, whose members are drawn from the councils of the National Academy of Sciences, the National Academy of Engineering, and the Institute of Medicine. The members of the committee responsible for the report were chosen for their special competences and with regard for appropriate balance.

This work was sponsored by the U.S. Department of Energy, Contract No. DE-FC01-94EW54069. All opinions, findings, conclusions, and recommendations expressed herein are those of the authors and do not necessarily reflect the views of the Department of Energy.

International Standard Book Number 0-309-06549-6 

Library of Congress Catalog Card Number 99-65127

Additional copies of this report are available from:

National Academy Press
2101 Constitution Ave., NW
Box 285
Washington, DC 20055
800-624-6242
202-334-3313 (in the Washington Metropolitan Area)
http://www.nap.edu

Printed in the United States of America

COMMITTEE ON TECHNOLOGIES FOR CLEANUP OF SUBSURFACE CONTAMINANTS IN THE DOE WEAPONS COMPLEX

C. HERB WARD, *Chair,* Rice University, Houston, Texas
HERBERT E. ALLEN, University of Delaware, Newark
RICHARD BELSEY, Physicians for Social Responsibility, Portland, Oregon
KIRK W. BROWN, Texas A&M University, College Station
RANDALL J. CHARBENEAU, University of Texas, Austin
RICHARD A. CONWAY, Union Carbide Corporation (retired), South Charleston, West Virginia
HELEN E. DAWSON, Colorado School of Mines, Golden
JOHN C. FOUNTAIN, State University of New York, Buffalo
RICHARD L. JOHNSON, Oregon Graduate Institute of Science and Technology, Portland
ROBERT D. NORRIS, Eckenfelder, Brown and Caldwell, Nashville, Tennessee
FREDERICK G. POHLAND, University of Pittsburgh, Pittsburgh, Pennsylvania
KARL K. TUREKIAN, Yale University, New Haven, Connecticut
JOHN C. WESTALL, Oregon State University, Corvallis

Staff

JACQUELINE A. MACDONALD, Study Director
SUSAN B. MOCKLER, Research Associate
LATRICIA C. BAILEY, Project Assistant
ERIKA L. WILLIAMS, Research Assistant

BOARD ON RADIOACTIVE WASTE MANAGEMENT

MICHAEL C. KAVANAUGH, *Chair*, Malcolm Pirnie, Inc., Oakland, California
JOHN F. AHEARNE, *Vice-Chair*, Sigma Xi, The Scientific Research Society, and Duke University, Research Triangle Park and Durham, North Carolina
ROBERT J. BUDNITZ, Future Resources Associates, Inc., Berkeley, California
ANDREW P. CAPUTO, Natural Resources Defense Council, Washington, D.C.
MARY R. ENGLISH, University of Tennessee, Knoxville
DARLEANE C. HOFFMAN, Lawrence Berkeley Laboratory, Berkeley, California
JAMES H. JOHNSON, JR., Howard University, Washington, D.C.
ROGER E. KASPERSON, Clark University, Worcester, Massachusetts
JAMES O. LECKIE, Stanford University, Stanford, California
JANE C. S. LONG, University of Nevada, Reno
CHARLES McCOMBIE, NAGRA, Wettingen, Switzerland
WILLIAM A. MILLS, Oak Ridge Associated Universities (retired), Olney, Maryland
D. WARNER NORTH, NorthWorks, Inc., Belmont, California
MARTIN J. STEINDLER, Argonne National Laboratory, Argonne, Illinois
JOHN J. TAYLOR, Electric Power Research Institute, Palo Alto, California
MARY LOU ZOBACK, U.S. Geological Survey, Menlo Park, California

NRC Staff

KEVIN D. CROWLEY, Director
ROBERT S. ANDREWS, Senior Staff Officer
THOMAS E. KIESS, Senior Staff Officer
JOHN R. WILEY, Senior Staff Officer
SUSAN B. MOCKLER, Research Associate
TONI GREENLEAF, Administrative Associate
MATTHEW BAXTER-PARROTT, Project Assistant
LATRICIA C. BAILEY, Project Assistant
PATRICIA A. JONES, Senior Project Assistant
LAURA D. LLANOS, Project Assistant
ANGELA R. TAYLOR, Senior Project Assistant

COMMISSION ON GEOSCIENCES, ENVIRONMENT, AND RESOURCES

The National Academy of Sciences is a private, nonprofit, self-perpetuating society of distinguished scholars engaged in scientific and engineering research, dedicated to the furtherance of science and technology and to their use for the general welfare. Upon the authority of the charter granted to it by the Congress in 1863, the Academy has a mandate that requires it to advise the federal government on scientific and technical matters. Dr. Bruce Alberts is president of the National Academy of Sciences.

The National Academy of Engineering was established in 1964, under the charter of the National Academy of Sciences, as a parallel organization of outstanding engineers. It is autonomous in its administration and in the selection of its members, sharing with the National Academy of Sciences the responsibility for advising the federal government. The National Academy of Engineering also sponsors engineering programs aimed at meeting national needs, encourages education and research, and recognizes the superior achievements of engineers. Dr. William A. Wulf is president of the National Academy of Engineering.

The Institute of Medicine was established in 1970 by the National Academy of Sciences to secure the services of eminent members of appropriate professions in the examination of policy matters pertaining to the health of the public. The Institute acts under the responsibility given to the National Academy of Sciences by its congressional charter to be an adviser to the federal government, and, upon its own initiative, to identify issues of medical care, research, and education. Dr. Kenneth I. Shine is president of the Institute of Medicine.

The National Research Council was organized by the National Academy of Sciences in 1916 to associate the broad community of science and technology with the Academy's purposes of furthering knowledge and advising the federal government. Functioning in accordance with general policies determined by the Academy, the Council has become the principal operating agency of both the National Academy of Sciences and the National Academy of Engineering in providing services to the government, the public, and the scientific and engineering communities. The Council is administered jointly by both Academies and the Institute of Medicine. Dr. Bruce Alberts and Dr. William A. Wulf are chairman and vice-chairman, respectively, of the National Research Council.

Preface

Environmental legislation resulting in the Resource Conservation and Recovery Act (RCRA) of 1976 and the Comprehensive Environmental Response, Compensation, and Liability Act (CERCLA, commonly known as Superfund) of 1980 led to the discovery of massive contamination of groundwater and soil at sites scattered across the United States. The original Superfund of $1.6 billion was based on an estimated average cost of $3.6 million per site for cleanup of 400 contaminated sites. However, Superfund was a new enterprise not based on past experience. By 1990, the Environmental Protection Agency (EPA) estimated a total cleanup cost of $27 billion at an average cost of $26 million per site. As the nation continued to gain experience in hazardous waste remediation, EPA estimated that the Superfund National Priorities List (NPL) could grow to more than 2,000 sites and that estimated costs could increase to the range of $100 billion to $500 billion. More recent estimates indicated that under scenarios requiring cleanup to stringent standards, costs could exceed $1 trillion when accounting for sites owned by the Department of Defense (DOD), the Department of Energy (DOE), and state governments, in addition to privately owned sites. This brief history shows that estimation of total costs of cleaning up contaminated sites is highly uncertain, if not impossible.

Most cost estimates to date have been based on the use of conventional and readily available remediation technologies. However, those involved with site remediation have gradually recognized that, regardless of cost, the technology does not exist to effectively manage the most recalcitrant contamination problems. These difficult problems include dense nonaqueous-phase liquids (DNAPLs), metals, and radionuclides in groundwater and soil. The National Research Council (NRC) addressed the complexities of groundwater remediation in its 1994 study *Alternatives for Ground Water Cleanup*, which identified the limitations of conventional remediation technologies and served to heighten focus on this problem. Today, 19 years after Congress responded to public concern about Love Canal by creating the CERCLA program, we are faced with a paradigm shift: a recognition that the most difficult contamination problems cannot be solved with conventional technology and that cleanup to health-based standards will not be possible at every site.

Recognizing that inadequate technology is a critical limiting factor in meeting federal cleanup standards, during the past decade EPA, DOD, and DOE began programs to develop new and innovative environmental remediation technologies. Each agency focused on technology development to solve its most pressing problems, some of which were unique to the agency but many of which (including DNAPLs) were common across the contaminated landscape. Development of completely new, more effective, and less costly cleanup technology proved to be difficult, expensive, and time consuming. Hence, numerous existing technologies were redesigned for environmental cleanup. An important example of retooling of existing bodies of science and technology is the adaptation of surfactant- and cosolvent-enhanced oil recovery methods (used in the petroleum industry) for the removal of nonaqueous-phase liquids (NAPLs), such as gasoline and chlorinated solvents, from aquifers. Another is the adaptation of extractive metallurgy technology for the removal of metal contaminants, such as lead, from soil.

As new or redesigned technologies became available, a new problem surfaced—the unwillingness of regulatory agencies and the cleanup community to embrace them. Most of the new technologies were considered unproven, and the risk of their use and potential failure was unacceptable. In the environmental technology development community this phenomenon became known as part of the "Valley of Death," symbolizing the failure of most remediation technologies to progress successfully from the research and development stage to full-scale implementation. That is, good technologies never reached the commercial stage because of real or perceived risks in using them. The NRC addressed this problem in the 1997 study *Innovations in Ground Water and Soil Cleanup: From Concept to Commercialization*.

In 1995, under the guidance of the Board on Radioactive Waste Management (BRWM), the NRC appointed the Committee on Environmental Management Technologies (CEMT) to advise DOE's Office of Science and Technology on its environmental remediation technology development program. Because of the great breadth of the technological issues involved in cleanup of the nation's nuclear legacy, subcommittees were formed to address specific environmental media, waste types, and technology areas. CEMT's two annual reports identified the need for in-depth review and analysis of technology development beyond the scope and charge of its subcommittees. As a result, the NRC formed several new committees in 1997 to advise DOE on specific areas of technology development. One of these committees was the Committee on Technologies for Cleanup of Subsurface Contaminants in the DOE Weapons Complex, which wrote this report. The committee's charge was to focus on the most recalcitrant problems remaining in groundwater and soil: DNAPLs, metals, and radionuclides.

A study of any one of these contaminant groups could have been challenging. Addressing all three in one report was a significant test of the committee's knowledge and breadth. Physical and chemical properties of contaminants determine their behavior in environmental media. Because of the diverse properties of DNAPLs, metals, and radionuclides, scientists and engineers seldom work with more than one of these groups. Regardless, our assignment was to review the status of DOE's subsurface remediation technology development program for all three groups and provide recommendations to help direct future activities. Understandably, all members of the committee were not able to contribute equally, but the diversity of backgrounds and knowledge that committee members were able to bring to this study provided for rich and intellectually challenging discussions that generally led to consensus. We hope our efforts will suffice to identify the current state of the art of technology development for remediation of these contaminant groups and that we provide insights that will prove useful to DOE and the nation.

This study was conducted by a very diverse and talented group of scientists and engineers. I am indebted to them for their hard work and dedication to our assignment. Most of us, I believe, may have learned more than we contributed. That is our reward. Studies of this depth and breadth, however, are beyond the ability of a committee to bring to completion on its own. A skilled and competent NRC staff is essential. We were blessed by having one of the NRC's most consummate professionals as our study director. Having Jackie MacDonald work with us was not a chance draw. I requested that she serve as our study director if her involvement in the project could be arranged. My past work with her has been very productive. She played a pivotal role in the *Alternatives for Ground Water Cleanup* study, helping us synthesize information on a highly complex, controversial, and politically charged issue. She was also study director of the highly insightful study *Innovations in Ground Water and Soil Cleanup*, which identified key issues limiting the development of new environmental remediation technology. I consider Jackie MacDonald's participation critical to the success of this study.

Several other NRC staff members also were essential to completion of this project. During the early part of the study, Rebecca Burka and Erika Williams managed logistical arrangements for committee meetings and helped with research. Latricia Bailey effectively took over this role for the later part of the study and also managed production of the report manuscript; her efficiency and attention to detail are greatly appreciated by the committee members. Susan Mockler contributed valuable research assistance and help in inviting appropriate guests to speak at committee meetings for the early part of the study.

The committee is also indebted to the many scientists and others from inside and outside DOE, too numerous to list here, who took the time to present information to the committee. Finally, we are most appreciative of the managers of DOE's Subsurface Contaminants Focus Area (SCFA) program for their cooperation in this study. SCFA staff members were extremely helpful in providing the committee with information needed to assess DOE's progress in developing new subsurface remediation technologies and in coordinating arrangements for several meetings at DOE installations. We are especially grateful to Jim Wright, Skip Chamberlain, Phil Washer, and Joan Baum for their cooperation and insights. Of course, this study would not have been possible without financial sponsorship from the Department of Energy.

Although this report focuses on contaminated sites owned by DOE, the information on remediation technologies and problems in cleanup applies well beyond facilities in the former nuclear weapons production complex. We hope that this report will help guide development of the next generation of remediation technologies for broad use nationwide.

This report has been reviewed in draft form by individuals chosen for their diverse perspectives and technical expertise, in accordance with procedures approved by the NRC's Report Review Committee. The purpose of this independent review is to provide candid and critical comments that will assist the insitution in making the published report as sound as possible and to ensure that the report meets institutional standards for objectivity, evidence, and responsiveness to the study charge. The review comments and draft manuscript remain confidential to protect the integrity of the deliberative process. We wish to thank the following individuals for their participation in the review of this report: Edgar Berkey, Concurrent Technologies Corporation; David Blowes, University of Waterloo; Suresh Chandra Rao, University of Florida; Roy E. Gephardt, Pacific Northwest National Laboratory; Walter Kovalick, Jr., U.S. Environmental

Protection Agency; Jane C. S. Long, University of Nevada; Richard A. Meserve, Covington & Burling; Dade Moeller, Dade Moeller & Associates, Inc.; and Philip A. Palmer, E.I. DuPont de Nemours & Company. While the individuals listed above have provided constructive comments and suggestions, it must be emphasized that responsibility for the final content of this report rests entirely with the authoring committee and the institution.

C. Herb Ward
Rice University
Houston, Texas

Contents

EXECUTIVE SUMMARY 1
DOE's Progress in Groundwater and Soil Remediation, 2
The Changing Regulatory Environment, 3
Technologies for Metals and Radionuclides, 4
Technologies for DNAPLs, 4
DOE Remediation Technology Development, 7
Recommendations, 11

1 INTRODUCTION: DOE'S GROUNDWATER AND SOIL CONTAMINATION PROBLEM 15
Limitations of Conventional Groundwater and Soil Cleanup Technologies, 17
DOE's Program for Developing Groundwater and Soil Cleanup Technologies, 19
Dimensions of DOE's Subsurface Contamination Problem, 21
DOE's Progress to Date in Cleaning Up Groundwater and Soil Contamination, 35

2 THE CHANGING REGULATORY ENVIRONMENT 39
Overview of Applicable Federal Regulations, 40
Baseline Cleanup Goals, 47
Changing Regulatory Environment, 54
Conclusions, 68

3 METALS AND RADIONUCLIDES: TECHNOLOGIES FOR CHARACTERIZATION, REMEDIATION, AND CONTAINMENT 72
Factors Affecting Risks of Metal and Radionuclide Contamination, 72
Geochemical Characteristics of Metal and Radionuclide Contaminants: Effects on Treatment Options, 76
Characterization of Metal and Radionuclide Contamination, 77
Physical Barriers for Containing Contaminants, 84
Technologies for Immobilizing Metals and Radionuclides, 96
Technologies for Mobilizing and Extracting Metals and Radionuclides, 112
Conclusions, 120

4 DNAPLS: TECHNOLOGIES FOR CHARACTERIZATION, REMEDIATION, AND CONTAINMENT 129
The DNAPL Problem, 129
Characterization of DNAPL Contamination, 133
Remediation Technologies for DNAPL Source Zones, 140
Remediation Technologies for Plumes of Dissolved DNAPL Contaminants, 171
Common Limitations of DNAPL Remediation Technologies, 192
Conclusions, 193

5 DOE REMEDIATION TECHNOLOGY DEVELOPMENT: PAST EXPERIENCE AND FUTURE DIRECTIONS 202
Barriers to Innovative Technology Use at DOE Sites, 203
DOE Steps to Increase Innovative Technology Deployment, 207
Deployment of Innovative Remediation Technologies at DOE Installations, 212
Effectiveness of Reforms in Promoting Deployments, 218
SCFA Technology Development Achievements, 220
Conclusions, 235

6 FINDINGS AND RECOMMENDATIONS 240
Setting Technology Development Priorities, 240
Improving Overall Program Direction, 244
Overcoming Barriers to Deployment, 245
Addressing Budget Limitations, 247

APPENDIXES

A Facilities at Which DOE Is Responsible for Environmental Cleanup 253
B SCFA Technology Deployment Report 264
C Biographical Sketches of Committee Members and Staff 269
D Acronyms 274

INDEX 277

Executive Summary

Cleaning up contamination at installations that were part of the former nuclear weapons production complex is the most costly environmental restoration project in U.S. history. The Department of Energy (DOE), which is responsible for these installations, has spent between $5.6 billion and $7.2 billion per year on environmental management over the past several years. Despite these expenditures, progress has been limited. Although management and institutional problems have slowed the cleanup effort, technical limitations also have played a role. Effective technologies do not exist for treating many of the common groundwater and soil contaminants at DOE facilities.

This report advises DOE on technologies and strategies for cleaning up three types of contaminants in groundwater and soil: (1) metals, (2) radionuclides, and (3) dense nonaqueous-phase liquids (DNAPLs), such as solvents used in manufacturing nuclear weapons components.[1] Metals and DNAPLs are common not only in the weapons complex but also at contaminated sites nationwide owned by other federal agencies and private companies. They have proven especially challenging to clean up, not just for DOE but also for others responsible for contaminated sites. Although the recommendations in this report are designed for DOE, the bulk of the report will

---

[1] As used in this report, "cleanup" means removing contaminant mass from groundwater or soil, immobilizing the contaminant in the ground to keep it from spreading, or containing the contamination in place

be useful to anyone involved in the cleanup of contaminated sites. The report contains reviews of regulations applicable to contaminated sites, the state of the art in remediation technology development, and obstacles to technology development that apply well beyond sites in the DOE weapons complex.

Within DOE, the Subsurface Contaminants Focus Area (SCFA) in the Office of Science and Technology is responsible for developing technologies to clean up metals, radionuclides, and DNAPLs in groundwater and soil. SCFA, like others involved in developing technologies to solve these problems, has encountered major obstacles. This report recommends where SCFA should direct its technology development program to achieve the most progress.

This report was prepared by the National Research Council's (NRC's) Committee on Technologies for Cleanup of Subsurface Contaminants in the DOE Weapons Complex. The NRC appointed this committee in 1997 at DOE's request. The committee included experts in hydrogeology, environmental engineering, geochemistry, soil science, and public health. Members were selected from academia, consulting firms, private industries, and public interest groups to represent a range of perspectives on DOE contamination problems. The committee's conclusions are based on a review of relevant technical literature, briefings by staff from DOE and environmental regulatory agencies, visits to several DOE installations, consultations with other experts, and the knowledge and experiences of committee members.

DOE'S PROGRESS IN GROUNDWATER AND SOIL REMEDIATION

In total, DOE is responsible for cleanup of 113 installations in 30 states. To date, DOE has identified approximately 10,000 individual contaminant release sites within these installations that contain groundwater and/or soil contamination; continuing investigations may uncover further contamination. Current estimates indicate that some 1.8×10^9 m^3 of groundwater and 75×10^6 m^3 of soil are affected. These contamination problems date from the start in 1942 of the Manhattan Project to develop nuclear weapons.

Assessing DOE's progress in cleaning up contaminated groundwater and soil is difficult because of data limitations, conflicting terminology, and lack of an agreed-upon metric for measuring success (see Chapter 1 for details). DOE's Office of Environmental Restoration reported that, as of 1998, remedies had been selected for 27 of 92 active groundwater cleanup projects and for 163 of 221 soil cleanup projects. Some of these projects include multiple contaminated sites, so it is unclear what percentage of the 10,000 contaminant release sites are being cleaned up. However, it appears that the number is small, and progress has been minimal.

DOE's attempts to clean up contaminated groundwater and soil have been limited in part by technological difficulties. Conventional pump-and-treat systems for contaminated groundwater, which are slated for use at the bulk of DOE sites where groundwater restoration is under way, often cannot achieve cleanup goals for many of the types of contamination scenarios encountered at DOE installations. For example, a 1994 NRC survey of 77 contaminated sites showed that pump-and-treat systems had achieved cleanup goals at just 8 of the sites. Excavation, the most common remedy for contaminated soil at DOE installations, can increase the risk of exposure to contamination (exactly the problem remediation is supposed to avoid) and destroy native ecosystems, and in many circumstances it is costly. Because of such limitations, new technologies are needed to enable DOE to achieve remediation requirements for groundwater and soil at reasonable cost.

THE CHANGING REGULATORY ENVIRONMENT

An essential part of planning SCFA's program to develop new remediation technologies is an understanding of what cleanup requirements DOE must achieve, because these determine the desired technology performance goals. Groundwater and soil restoration goals have not yet been specified for many DOE sites, making it difficult to establish technology performance goals. Nonetheless, when these goals are established they must satisfy the requirements of applicable regulations: generally the Resource Conservation and Recovery Act (RCRA); Comprehensive Environmental Response, Compensation, and Liability Act (CERCLA); Uranium Mill Tailings Remediation Control Act (UMTRCA); or a combination of these.

Historically, regulations under these laws have required that at most sites DOE restore contaminated groundwater to drinking water standards, known as maximum contaminant levels (MCLs), or to special standards designed specifically for UMTRCA sites. Regulators at DOE sites usually require that soil cleanup meet specifications outlined in a soil screening guidance document developed by the Environmental Protection Agency (EPA). In general, DOE must achieve groundwater and soil cleanup standards across the full site, except in specially designated waste management areas where remaining contaminants will be contained in place.

In the past few years, changes in baseline cleanup standards for groundwater and soil and in the overall process of site cleanup have become increasingly common, in part due to technical limitations and costs. Changes include increases in the number of waivers to baseline cleanup standards and to original site remedies, increasing use of natural attenuation in place of engineered remedies, emergence of brownfields programs with less stringent cleanup standards, and emergence of new risk-based methods for priority setting (all described in detail in Chapter 2). These new paradigms may affect the selection of cleanup goals for DOE sites and, correspondingly, the suite of possible remediation technologies for achieving those goals. Nonetheless, SCFA will have to continue developing technologies capable of cleaning up difficult sites with long-term liability concerns and of meeting baseline standards at the many sites where these will remain as cleanup goals.

TECHNOLOGIES FOR METALS AND RADIONUCLIDES

Cleanup of metals and radionuclides in the subsurface is complicated by a number of factors. Metals and radionuclides have multiple possible oxidation states with different mobilities, can partition to organic matter present in soil, can sorb to other soil components, and can precipitate. All of these factors can affect the performance of remediation technologies. Few well-established technologies are available for treating these types of contaminants, but a number of promising technologies are in the development stage.

Because metal and radionuclide contaminants are generally nondegradable, treatment technologies must involve some form of mobilization of the contaminant (in order to move it to a location where it can be treated) or immobilization (in order to stabilize it in place and prevent further spreading). Table ES-1 lists established and emerging technologies for mobilization, followed by treatment, or immobilization of metals and radionuclides (see Chapter 3 for details). As is clear from the table, additional work is needed to increase the range of proven options for treating metals and radionuclides in situ and for extracting them (without excavation) for ex situ treatment; most of the technologies listed in the table are still in the development stage.

TECHNOLOGIES FOR DNAPLS

Conventional technologies are generally ineffective at restoring DNAPL-contaminated sites, as has been well documented in previous studies by the NRC and others. Chlorinated solvents are the

TABLE ES-1 Technologies for Remediation of Metals and Radionuclides in Groundwater and Soil

Technology	Applicability
Subsurface barriers	Well-established method for preventing the spread of metal and radionuclide contaminants in groundwater. Vertical barriers are widely available; methods are being developed for installation of horizontal barriers beneath existing waste.
In situ vitrification	Developing technology for immobilizing metal and radionuclide contaminants in the subsurface. It is particularly suitable for sites with high concentrations of long-lived radioisotopes within 6 to 9 m of the soil surface (depending on water table depth and soil moisture). This technology may be able to treat mixtures of organic and inorganic contaminants. However, it is among the most expensive of treatment options.
Solidification and stabilization	Mature technologies for ex situ immobilization of contaminated soil. Less well developed for use in situ because of the difficulty of ensuring sufficient mixing. Improved mixing methods are being tested.
Permeable reactive barriers	Among the most promising and rapidly developing treatment technologies for treating metals, radionuclides, and mixtures of organic and inorganic contaminants. These barriers either intercept the flow of contaminated groundwater with a subsurface zone in which reactive materials have been installed to treat the contaminants or direct water flow through such a zone; a variety of reactive materials have been tested successfully. Operation and maintenance costs are relatively low because little or no energy input is required to maintain the system. Because the technology is relatively new, the longevity of reactive materials is a major uncertainty.
In situ redox manipulation	A developing method for treating metals and radionuclides at depths at which digging the trenches required for barrier technologies is impractical. The technology involves injection of chemical reductants into the ground to create reducing conditions that lead to immobilization of certain metals and radionuclides. It is especially well suited for elements (such as chromium) that can be reduced to solids that are resistant to reoxidation by ambient oxygen. It is less suitable for elements (such as technetium) that reoxidize easily. As with reactive barriers, the longevity of the treated zone is unknown.

continues on next page

TABLE ES-1 Continued

Technology	Applicability
Bioremediation	Developing method using subsurface microorganisms to mobilize or immobilize metals and radionuclides. If further developed, the technology may be able to treat combinations of organic and inorganic contaminants at relatively low cost and with relatively little disruption to the site.
Electrokinetic systems	Developing technologies in which an electric field is applied to soil either to stabilize the contaminants in situ or to mobilize them for extraction near the electrodes. Extensive field tests of electrokinetic systems for the remediation of metal and radionuclide contamination have yet to be conducted in the United States. If better developed, the method would be appropriate for treating media with very low hydraulic conductivities.
Soil washing	Established technology for the ex situ separation of fine-grained soils, which generally harbor most of the contamination, from coarser soils. Because this is an ex situ process, it requires excavation of the soils and has all the limitations imposed by excavation.
Soil flushing	Developing technology for treating metals and radionuclides in situ by flushing contaminated soils with solutions designed to recover the contaminants. This technology is derived from the mining industry but has not yet been widely applied for environmental remediation of metals and radionuclides.
Phytoremediation	Developing technology in which specially selected or engineered plant species are grown and harvested after taking up metals and radionuclides through their roots. Phytoremediation has been field tested for treating a range of metals and radionuclides. It is most applicable to large areas of surface soils with low to moderate levels of contamination. Costs are low, and implementation is relatively easy, but mobilization of contaminants and transport to the groundwater is a risk when certain soil amendments are used to facilitate plant uptake of the contaminants.

predominant DNAPL contaminants at DOE sites. These solvents have low solubilities in water and are denser than water. They tend to remain as a separate organic liquid in the subsurface, rather than mixing with water. A portion of a DNAPL contaminant will become entrapped in soil pores, while the rest sinks beneath the water table. Small amounts of separate-phase solvent can then dissolve in the flowing groundwater at levels high enough to make the water unsafe for drinking.

Solutions to DNAPL contamination problems are best approached by dividing the problem into two distinct elements: (1) the DNAPL source zone, consisting of areas of the subsurface containing undissolved solvents entrapped in soil pores or traveling separately from the water, and (2) the dissolved plume, consisting of water that has been contaminated by components of DNAPLs that have dissolved. Several emerging technologies have shown the ability to remove mass relatively rapidly from DNAPL source zones. Other innovative technologies have demonstrated the ability to clean up plumes of dissolved contaminants.

Table ES-2 summarizes technologies for treating DNAPL source zones and dissolved plumes emanating from DNAPL sources (see Chapter 4 for details). Although these technologies show promise, determining the ultimate level of cleanup attainable for each is not possible because of the lack of carefully controlled field tests. Each of the technologies is based on well-established chemical and physical principles and thus is more likely to be limited by hydrogeologic conditions (especially geological heterogeneities, which can interfere with circulation of treatment fluids and water or can limit access to the subsurface) than by limitations of the processes themselves. Nonetheless, more field tests are needed to demonstrate performance levels under a variety of hydrogeologic conditions.

DOE REMEDIATION TECHNOLOGY DEVELOPMENT

SCFA has helped to develop a number of innovative technologies for remediation of metals, radionuclides, and DNAPLs, but use of these technologies in actual DOE cleanups has been limited (see Chapter 5). For example, the Office of Environmental Restoration reported that the predominant remedies for groundwater contamination are conventional pump-and-treat systems (used at 41 percent of sites) and natural attenuation (used at 22 percent of sites). Further, no-action alternatives are being used more often than any one innovative technology. The environmental restoration office reported two uses each of air sparging and free product recovery systems and one use each of

TABLE ES-2 Technologies for Remediation of DNAPLs in Groundwater and Soil

Technology	Applicability
Soil vapor extraction	Effective at cleaning up source zones containing volatile compounds in homogeneous, permeable soils; with addition of thermal processes, the technology can be extended to semivolatile compounds. Thorough removal of DNAPLs requires sufficient flow through the entire source zone, which may be difficult to achieve.
Steam	Demonstrated ability to clean up DNAPL source areas in permeable soil in both the saturated and the unsaturated zones. It may be combined with electrical heating when finer-grained layers are present. Heterogeneities in geologic materials in the subsurface may limit efficiency of this process.
Surfactant flooding	Demonstrated to effectively remove large masses of nonaqueous-phase liquids from source zones in permeable aquifers. Geologic heterogeneities and nonuniform contaminant distribution may reduce the efficiency of this process.
Cosolvent flooding	Has shown potential for solubilizing large masses of nonaqueous-phase liquids. Geologic heterogeneities and nonuniform contaminant distribution may reduce process efficiency.
In situ oxidation	Proven to be effective at destruction of specific chlorinated DNAPL compounds in source zones in permeable, relatively homogeneous soils. Geologic heterogeneities may reduce the efficiency of these processes, and mass transfer limitations may limit the volume of DNAPL that can be treated efficiently.
Electrical heating and electrokinetic methods	Have shown potential for remediation of dissolved contaminants from DNAPLs in low-permeability units. Significant data are available from field trials of electrical heating systems, but data are inadequate to verify the effectiveness of electrokinetic methods for treating DNAPL source zones.
Bioremediation	Demonstrated method for stimulating microorganisms in the subsurface to degrade chlorinated compounds. Degradation takes place primarily in the dissolved phase. Treatment of DNAPL source zones using biodegradation methods probably is not practical because of the long time required for dissolution.

continues on next page

TABLE ES-2 Continued

Technology	Applicability
Phytoremediation	Emerging method that uses plants to enhance microbial degradation of contaminants, take up contaminants, or provide hydraulic containment. Results are not yet conclusive for application to dissolved contaminants from DNAPLs, but field tests are under way.
In situ vitrification	Demonstrated as effective for converting soil to a molten material that solidifies upon cooling and for producing temperatures that should lead to the destruction or mobilization of DNAPL compounds. However, data on the performance of this technology at DNAPL sites are insufficient to provide a meaningful evaluation at this time.
Reactive barrier walls	Have shown great promise for treatment of dissolved plumes of contamination from chlorinated solvents. Although these technologies do not directly clean the DNAPL source zones, they limit the migration of plumes of contamination emanating from these zones. Uncertainties over the longevity of barrier walls are among the main limitations of this technology.

thermally enhanced vapor extraction systems and passive reactive barriers. Excavation, followed by ex situ treatment or disposal, is still the predominant remedy for contaminated soil.

The major barrier to deployment of SCFA's technologies is lack of demand from individual DOE cleanup operations (the end users of SCFA technologies). Other factors that have interfered with deployment of SCFA's technologies include regulatory requirements that favor conventional technologies, inconsistencies in technology selection processes and cleanup goals, and SCFA budget limitations.

The demand for innovative remediation technologies at DOE installations is lagging. In part, demand is lacking because incentives for rapid, cost-effective cleanup of DOE installations are inadequate. On the contrary, rapid cleanup of DOE sites can lead to loss of revenue for the contractor responsible for managing cleanup at the site and loss of local jobs once the cleanup is completed and the site closed. Further, DOE site managers can hesitate to approve the use of innovative remediation technologies due to the risk that if the technology fails, they will still be liable for paying for the cleanup.

Also limiting demand for SCFA technologies is insufficient involvement of technology end users in setting SCFA's technology development priorities. End users have to be involved in determining whether to continue funding for specific projects and in ensuring that the technologies being developed meet site needs. SCFA also must provide these end users with adequate technical support for implementing new technologies. Unless SCFA can better connect its R&D effort with technology end users to first set the R&D direction and then work cooperatively with them to employ the technologies in specific applications, SCFA expenditures will continue to show modest results. SCFA has initiated strategies to increase end user involvement, but in fiscal year 1998 it was unable to implement these new strategies because the entire SCFA budget went to paying for projects that began before SCFA was formed.

Regulatory problems have also interfered with deployment of innovative remediation technologies at DOE installations. Especially problematic are the slow, linear nature of the regulatory process and inconsistencies in the way the process is applied from site to site. These regulatory problems can delay the selection of remediation technologies (which further reduces demand) and result in the use of outdated technologies chosen years before site cleanup begins (although at some sites regulators allow changes to the original cleanup plans). Regulatory inconsistencies create uncertainties about whether a technology proven at one location will meet the regulatory requirements at another location, making contractors hesitant to take the risk of using an innovative technology.

The U.S. General Accounting Office (GAO) has, in past reports, pointed to management problems in the Office of Science and Technology as another reason for the limited success of DOE's technology development programs. For instance, in reviews in 1992 through 1994, the GAO determined that the Office of Science and Technology lacked sufficient mechanisms for eliminating poorly performing projects, performing comprehensive assessments of technology needs, and preventing overlap in technology development work. The Office of Science and Technology has instituted several management reforms to address these problems.

Large budget swings are a final factor that has contributed to the difficulties of SCFA's program. SCFA's budget has been cut substantially: from a high of $82.1 million in 1994 to a level of $14.7 million in fiscal year 1998, of which $5 million was earmarked by Congress, leaving SCFA with a budget of $9.7 million. The fiscal year 1999 budget of $25 million, although an increase over the 1998 level, is approximately equal to the average price of cleaning up a single CERCLA site. The current budget allows only a limited number of technology development projects to go forward and may not be sufficient for the large field demonstrations needed to advance new technologies.

Despite the slow progress in deploying innovative remediation technologies at DOE installations, SCFA has helped to develop a number of technologies that have shown considerable promise. Notable SCFA accomplishments in developing systems for remediation of metals and radionuclides include work on in situ redox manipulation for chromium contamination at Hanford, horizontal barriers for waste containment at the Idaho National Engineering and Environmental Laboratory, and penetrometer systems for characterizing metals and radionuclides in the subsurface. Achievements in the development of systems for remediation of DNAPLs include work on steam technologies at Lawrence Livermore National Laboratory, electrical resistance heating to enhance recovery of DNAPLs by soil vapor extraction in low-permeability soils at several DOE installations, and collaborative work with private industries to develop and field test electrokinetic systems for DNAPL remediation. These successful projects, described in detail in Chapter 5, can provide models for future SCFA work.

RECOMMENDATIONS

SCFA has an important mission to fulfill in developing technologies for cleanup of metals, radionuclides, and DNAPLs in the subsurface. SCFA's past success in developing technologies that are later deployed in the field has been limited by a number of factors, including lack of customer demand, inadequate involvement of technology users in setting SCFA program priorities, regulatory obstacles, and budget limitations. Although some of these problems must be addressed by higher levels of DOE management, SCFA can take steps to increase the likelihood that the new technologies it helps develop will be deployed and to focus its financial resources on the most promising technologies. The committee developed recommendations to help improve the SCFA program in a variety of areas. Chapter 6 describes all of the recommendations in detail. Following are the highest priorities:

Setting Technology Development Priorities

- In situ remediation technologies should receive a higher priority in SCFA because of their potential to reduce exposure risks and costs.
- SCFA should fund tests designed to develop and determine performance limits for technologies capable of treating the types of contaminant mixtures that occur at DOE sites.

• SCFA should focus a portion of the program's work on development of remedial alternatives (including containment systems) that prevent migration of contaminants at sites where contaminant source areas cannot be treated. Methods for monitoring long-term performance of these systems should be included in this work.

Improving Overall Program Direction

• SCFA should continue its efforts to work more closely with technology end users in setting its overall program direction. Working with end users, SCFA should identify key technical gaps and prepare a national plan for developing technologies to fill these gaps. Although SCFA consulted with end users and developed a prioritized list of problem areas (known as work packages) for funding in fiscal year 1998, it was unable to use this list to guide its program because the entire SCFA budget went to supporting multiyear projects that began before SCFA was formed.

• SCFA should strive to increase the involvement of technology end users in planning the technology demonstrations it funds. End users should be involved in planning every demonstration that SCFA funds, as in the Accelerated Site Technology Deployment Program.

• SCFA should significantly increase use of peer review for (1) determining technology needs and (2) evaluating projects proposed for funding (see NRC, 1998, for guidelines on peer review). Peer reviews should carry sufficient weight to affect program funding.

• SCFA should improve the accuracy of its reporting of technology deployments. SCFA should use a consistent definition of deployment and should work with the Office of Environmental Restoration to verify the accuracy of its deployment report.

Overcoming Barriers to Deployment

• SCFA should sponsor more field demonstrations, such as those funded under the Accelerated Site Technology Deployment Program, to obtain credible performance and cost data. SCFA should consider whether sponsorship could include partial reimbursement for failed demonstrations, if an alternate remediation system has to be constructed to replace the failed one.

• SCFA should ensure that the project reports it provides contain enough technical information to evaluate potential technology performance and effectiveness relative to other technologies. The project descriptions contained in SCFA's periodic technology summary reports are not sufficiently detailed to serve this purpose. SCFA's project reports should follow the guidelines in the Federal Remediation Technologies Roundtable's *Guide to Documenting and Managing Cost and Performance Information for Remediation Projects* (FRTR, 1998).

- A key future role for the SCFA should be the development of design manuals for technologies that could be widely used across the weapons complex. Possible models include the Air Force Center for Environmental Excellence design manual for bioventing, the American Academy of Environmental Engineers WASTECH monograph series, and the Advanced Applied Technology Demonstration Facility surfactant-cosolvent manual.
- Appropriately qualified SCFA staff members (with in-depth knowledge of remediation technologies) should be available to serve as consultants on innovative technologies for DOE's environmental restoration program. These staff members also should develop periodic advisories for project managers on new, widely applicable technologies.

Addressing Budget Limitations

- DOE managers should reassess the priority of subsurface cleanup relative to other problems and, if the risk is sufficiently high, should increase remediation technology development funding accordingly.
- SCFA should pursue a variety of strategies to leverage its funding. Strategies include (1) improving collaborations with external technology developers to avoid duplication of their work, (2) developing closer ties with the Environmental Management Science Program, and (3) continuing involvement with working groups of the Remediation Technologies Development Forum.

In summary, DOE faces the challenge of cleaning up massive quantities of contaminated groundwater and soil with a suite of baseline technologies that are not adequate for the job. Although recent DOE budget projections have indicated that most groundwater at DOE installations will not be cleaned up, federal law requires groundwater cleanup, and political pressure to meet the federal requirements continues. DOE will thus have to continue to invest in developing groundwater and soil remediation technologies. As shown in Tables ES-1 and ES-2, a variety of emerging technologies for treating contaminated groundwater and soil are in the pipeline. DOE has to ensure that SCFA is adequately organized and supported to advance these types of technologies and to develop new technologies for contamination problems that still cannot be solved.

REFERENCES

FRTR (Federal Remediation Technologies Roundtable). 1998. Guide to Documenting and Managing Cost and Performance Information for Remediation Projects. Washington, D.C.: EPA.

NRC. 1998. Peer Review in Environmental Technology Development Programs: The Department of Energy's Office of Science and Technology. Washington, D.C.: National Academy Press.

1

Introduction: DOE's Groundwater and Soil Contamination Problem

The Department of Energy (DOE) faces monumental challenges in restoring the environment at installations that were part of the U.S. nuclear weapons production complex. Cleaning up these installations is the most costly environmental restoration project in U.S. history (Harden, 1996). DOE has spent between $5.6 billion and $7.2 billion per year on environmental management over the past several years for decontamination and decommissioning of nuclear reactors and other facilities, characterization of the types and locations of contaminants in the environment, and stabilization or removal of contaminants (GAO, 1997; Betts, 1998). The department projects that environmental management activities between now and 2070 will cost a total of $147.3 billion in 1998 dollars (DOE, 1998).

One important component of DOE's environmental management problem is the cleanup of groundwater and soil that were contaminated as a result of the range of activities associated with nuclear weapons production. Plumes of contaminated groundwater totaling an estimated 1.8×10^9 m^3 are migrating beneath DOE facilities, and an estimated 75×10^6 m^3 of soil are contaminated (DOE, 1997). DOE estimates that remediation of these resources will cost more than $15 billion in 1998 dollars (DOE, 1998).

Despite the large amount invested in DOE environmental management, progress on groundwater and soil remediation has been slow. Cleanup of most groundwater and soil contamination sites is in the early stages (EPA, 1997). Nontechnical factors—including management problems, inadequate incentives for DOE contractors, and regulatory obstacles—have contributed to the slow pace of groundwater and soil cleanup at DOE sites and are reviewed briefly in Chapter 5 of this report. However, technical problems also have limited DOE's progress and are the principal focus of this report. Technologies for remedying many of the types of soil and groundwater contamination problems found at DOE facilities are in the early stages of development.

This report focuses on three key categories of contaminants commonly found in soil and groundwater at DOE installations: (1) metals, (2) radionuclides, and (3) dense nonaqueous-phase liquids (DNAPLs), which are oily liquids that are denser than water. The report evaluates the technical options available for cleaning up these classes of contaminants. It also assesses DOE's programs for developing new remediation technologies to address these problems. Although the recommendations in the report are designed for DOE, the bulk of the information will be useful well beyond DOE. DNAPLs and metals are common contaminant classes at all contaminated sites, not just those owned by DOE.

This report was prepared by the National Research Council's (NRC's) Committee on Technologies for Cleanup of Subsurface Contaminants in the DOE Weapons Complex. The NRC appointed the committee in 1997 at the request of DOE to review technologies for characterizing, containing, and cleaning up metals, radionuclides, and DNAPLs in groundwater and soil. The committee included experts in hydrogeology, environmental engineering, geochemistry, soil science, and public health from academia, consulting firms, private industries, and public interest groups. During the course of its two-year study, the committee met six times to gather information and prepare this report. The committee also visited cleanup managers at three DOE installations: Lawrence Livermore National Laboratory in Livermore, California; the Savannah River Site in Aiken, South Carolina; and the Hanford Site in Richland, Washington. The committee's conclusions are based on a review of relevant technical literature; the expertise of committee members; and briefings to the committee by DOE managers, site cleanup contractors, Environmental Protection Agency (EPA) staff, and experts in site cleanup technologies from academia, federal laboratories, consulting firms, and industries.

DOE asked the committee to address five specific tasks:

1. identify and evaluate the complexity of subsurface conditions and contamination, focusing on metals, radionuclides, and DNAPLs, at selected DOE sites with geologic and hydrologic conditions that are representative of other sites across the weapons complex;
2. review and assess current EPA metal, radionuclide, and DNAPL

remediation guidelines, including risk-based end points, in reference to assessment of developing technologies;

3. review and assess developing technologies for application to characterization, containment, and cleanup of subsurface metal, radionuclide, and DNAPL contamination;

4. describe areas of uncertainty in the identified technologies; and

5. provide recommendations, as appropriate, on applications of subsurface remediation technologies for metals, radionuclides, and DNAPLs.

In addition to these tasks, which primarily involve a technical evaluation of remediation technologies and the performance standards they must meet, the committee conducted a limited review of DOE's program for developing new subsurface remediation technologies. This program is critical for ensuring that effective technologies are in the pipeline for addressing DOE groundwater and soil contamination problems that existing technologies cannot resolve.

This chapter outlines the magnitude of the groundwater and soil contamination problem at DOE facilities and briefly describes risks posed by this contamination, as currently understood. Understanding the nature of the problem is the first step in developing solutions; thus, this chapter provides an important context for understanding the technical evaluations in the later chapters of the report. Chapter 2 reviews the required cleanup goals for groundwater and soil contamination at DOE installations. Understanding these goals is important because they determine the performance standards, or "end states," that remediation technologies must achieve. Chapters 3 and 4 provide the bulk of the technical review in this report. Chapter 3 assesses the availability of technologies for characterization, remediation, and containment of radionuclides and metals in the subsurface, and Chapter 4 provides a similar assessment for DNAPLs. Chapter 5 evaluates the success of DOE's efforts to develop and deploy new technologies for metal, radionuclide, and DNAPL remediation and recommends future directions for DOE work in this area. Chapter 6 recommends strategies to improve DOE's program for developing groundwater and soil remediation technologies.

LIMITATIONS OF CONVENTIONAL GROUNDWATER AND SOIL CLEANUP TECHNOLOGIES

The limitations of conventional technologies for cleaning up contaminated groundwater and soil, whether at DOE installations or elsewhere, are now widely known among those involved in environmental restoration (NRC, 1994, 1997).

The conventional method for cleaning up contaminated groundwater is called "pumping and treating." Pump-and-treat systems operate by pumping large amounts of contaminated water from the subsurface via a series of wells, treating the water at the surface to remove contaminants, and then either reinjecting the water underground through a second set of wells or disposing of the water offsite. At large contaminated sites being cleaned up under the Comprehensive Environmental Response, Compensation, and Liability Act (CERCLA, also known as "Superfund"), this is still the predominant remedy, being used as the sole cleanup technology at 89 percent of sites with groundwater contamination (EPA, 1998). However, as has now been widely documented, these systems are often ineffective in restoring contaminated groundwater to regulatory standards because the flushing action created by pump-and-treat systems often is not sufficient to dislodge all of the contamination from the subsurface (NRC, 1994; MacDonald and Kavanaugh, 1994). Contaminants may diffuse into inaccessible regions of the subsurface or adhere to subsurface geologic materials. Small globules of DNAPL contaminants may become entrapped in the porous materials of the subsurface. The physical heterogeneity of the subsurface and the difficulties in characterizing this heterogeneity complicate delivery of treatment fluids to contaminated areas. All of these factors limit the ability to remove contaminants from the subsurface with pump-and-treat systems. In a 1994 review of pump-and-treat systems at 77 sites, the NRC found that cleanup goals had been achieved at 8 of the sites and were highly unlikely to be achieved at 34 of them (NRC, 1994). As discussed in more detail in Chapter 5, pump-and-treat systems, despite their limitations, are the predominant remedy at DOE sites where active cleanup is under way under the CERCLA program. Without the development of new technologies, then, it is highly unlikely that DOE cleanups will achieve regulatory standards for contaminated groundwater.

The conventional method for cleaning up contaminated soil is to excavate the soil and then either treat it to remove the contaminants or dispose of it in a specially designed landfill. Often, the treatment involves incineration. Although excavation removes contamination from the area of interest, there are major problems with the method. First, excavation can temporarily increase the risk of human exposure to contamination, both for site workers and for nearby residents who may be exposed to fugitive dusts. Second, excavation destroys the native ecosystem. Plants may be unable to grow unless new topsoil is added to the site after excavation. Third, treatment of excavated soil often involves incineration, and the public often objects to incineration because of the perceived potential for release of hazardous air pollutants when the soil is combusted (NRC, 1997). Fourth, digging up and disposing of tons of soil can be costly at sites where excavation is difficult, off-gas treatment is required, special health and safety measures are needed to protect workers, or the soil requires special disposal. As described in Chapter 5, excavation is the leading remedy being used to clean up soil at DOE's CERCLA sites. Development of new technologies could significantly reduce DOE's soil cleanup expenses and help to avoid problems associated with the destruction of native ecosystems and incineration.

DOE'S PROGRAM FOR DEVELOPING GROUNDWATER AND SOIL CLEANUP TECHNOLOGIES

DOE's Office of Environmental Management, which is responsible for overseeing cleanup at all of the department's contaminated installations, has long recognized the limitations of conventional technologies for cleaning up contaminated groundwater and soil, as well as for addressing other environmental concerns at DOE sites. Recognizing these technological limitations, the Office of Environmental Management in 1989 established the Office of Technology Development to develop technologies for DOE contamination problems for which good technical solutions are lacking. This office was later renamed the Office of Science and Technology (OST) and given expanded responsibilities. As the unique challenges posed by groundwater and soil cleanup became apparent, OST established a division devoted solely to the development of groundwater and soil cleanup technology. This division is now known as the Subsurface Contaminants Focus Area (SCFA). Because SCFA is the only unit within DOE with the primary mission of developing better solutions for contaminated groundwater and soil, the technical assessments and recommendations in this report are particularly relevant to SCFA.

SCFA prioritizes and provides funding for technology development efforts concerning containment of buried wastes and remediation of groundwater and soil contamination. DOE's Savannah River Site in Aiken, South Carolina, is responsible for administering the SCFA program. SCFA groups its technology development projects into four categories, known as "product lines": (1) source-term containment, (2) DNAPL remediation, (3) source-term remediation, and (4) metals and radionuclides. (A listing of projects currently funded under these product lines can be found at *http://www.envnet.org/scfa*.)

SCFA's budget has been cut in recent years, reflecting congres-

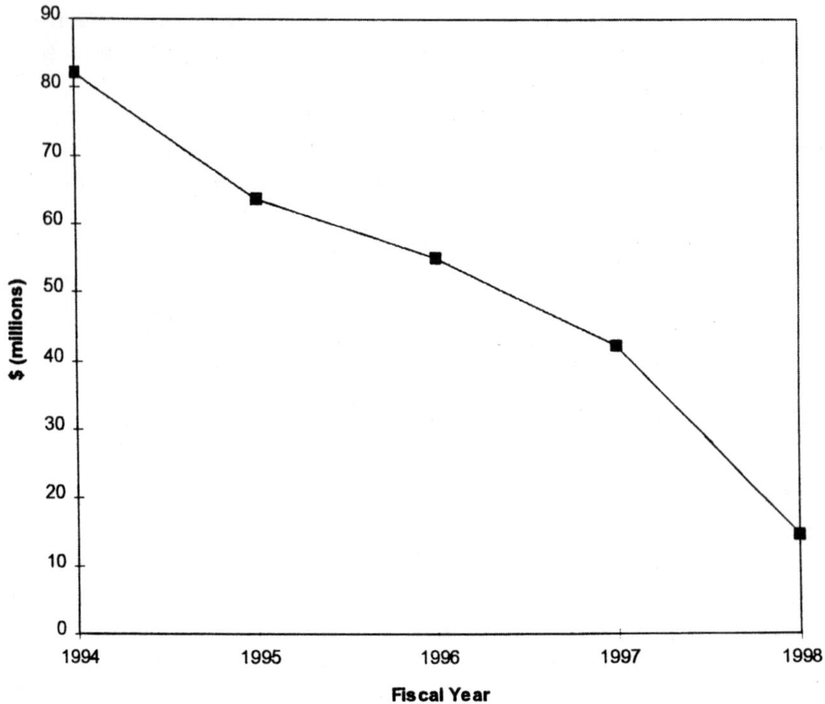

FIGURE 1-1 The SCFA budget over time. SOURCE: Budget data provided by SCFA.

sional dissatisfaction with the OST program as a whole (described in detail in Chapter 5). Congress cut the OST budget from a high of $410 million in 1995 to $274 million in 1998. SCFA's budget was cut from a high of $82.1 million in 1994 to a level of $14.7 million in fiscal year 1998 (see Figure 1-1). Congress earmarked $5 million of the $14.7 million appropriated in 1998, effectively leaving SCFA with a budget of $9.7 million. The earmarked funds were directed to the Western Energy Technology Center in Butte, Montana.

In the field of hazardous waste site cleanup, SCFA's 1998 budget of $9.7 million is a very small amount. Cleanup of a single private-sector CERCLA site costs an average of $24.7 million (CBO, 1994). Recent DOE cost projections have estimated that between 1997 and 2070, the department will spend $15 billion on cleanup of contaminant "release sites" (areas where contaminants were released and subsequently infiltrated soil and, often, groundwater) (DOE, 1998). This amount converts to annual expenses of approximately $770 million, when a discount rate of 5 percent is assumed. SCFA's 1998 budget represents about 1 percent of this spending. In fiscal year 1999, SCFA's budget was boosted to $25 million, but this is still a relatively small amount compared to the average cost of cleaning up a site. Further, the large budget swings have interfered with program planning.

DIMENSIONS OF DOE'S SUBSURFACE CONTAMINATION PROBLEM

Understanding the locations, types, and risks of contaminants present in the DOE weapons complex is the first step in determining remediation technology development needs. Whether a given process will be effective in cleaning up subsuface contamination at a specific site depends on the hydrogeology of the site, the characteristics of the contaminants, and the acceptable risk levels for the site. As described below, DOE's information on these dimensions of its subsurface contamination problems is incomplete.

Locations of DOE Facilities

Figure 1-2 shows the locations of DOE installations and other facilities at which DOE is responsible for environmental cleanup. Appendix A lists these facilities and their roles in nuclear weapons production. In total, DOE is charged with cleanup of 113 installations in 30 states (Probst and McGovern, 1998). DOE has identified approximately 10,000 individual contaminated sites within these facilities; continuing investigations may reveal further contamination (EPA, 1997).

Five of the installations shown on Figure 1-2 account for the majority (64 percent) of DOE's total projected costs for cleanup (EPA, 1997). These installations are the Rocky Flats Environmental Technology Site, the Idaho National Engineering and Environmental Laboratory, the Savannah River Site, the Oak Ridge Reservation, and the Hanford Site. These five facilities were essentially massive factories involved in nearly every phase of nuclear weapons production, from nuclear materials processing to weapons assembly (CERE, 1995). Table 1-1 shows the estimated volume of groundwater, soil, and sediment contamination at these major facilities. (These estimates are likely to change as DOE continues work to characterize its contaminated sites.) Box 1-1 describes the activities that led to environmental contamination at DOE installations.

In addition to cleaning up these major installations, DOE is responsible for cleaning up a large number of other facilities—some

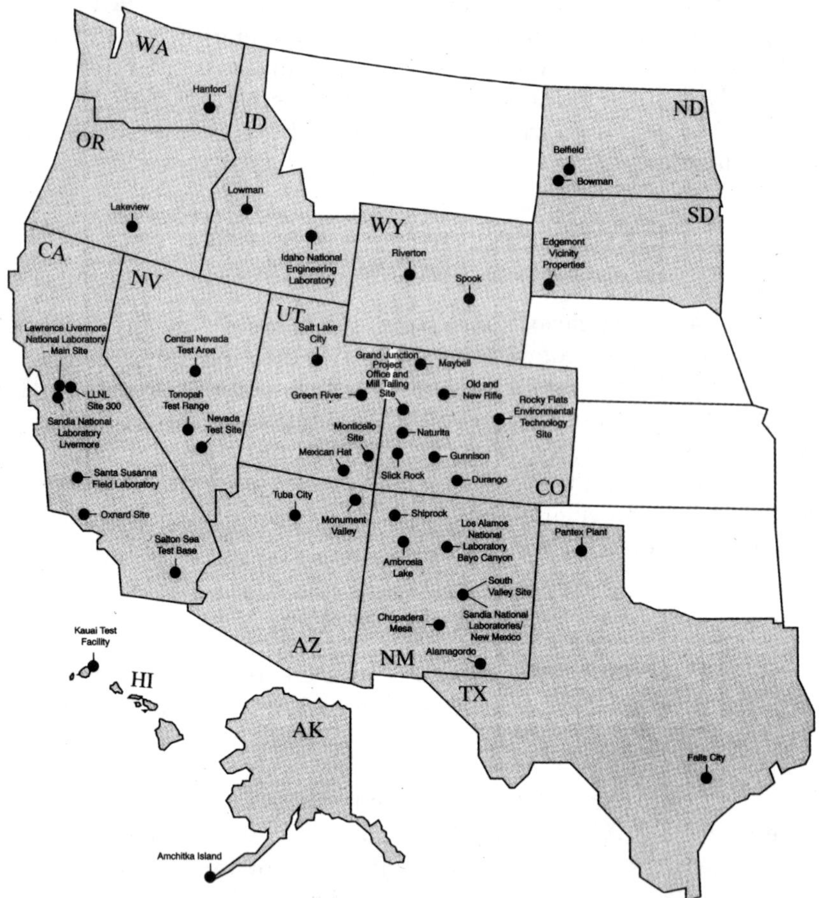

FIGURE 1-2 Contaminated facilities in the DOE complex. SOURCE: DOE, 1997.

owned by DOE and some not— that played smaller roles in the nuclear weapons production process. These other facilities include key DOE research laboratories, such as Los Alamos and Lawrence Livermore, and also a number of smaller operations that at one time or another were used for the processing of nuclear weapons materials. Twenty-four of the installations are former uranium processing facilities where DOE is cleaning up mine tailings and residual groundwater and soil contamination; these operations are part of what is known as the Uranium Mill Tailings Remediation Control Act (UMTRCA) project (EPA, 1997).

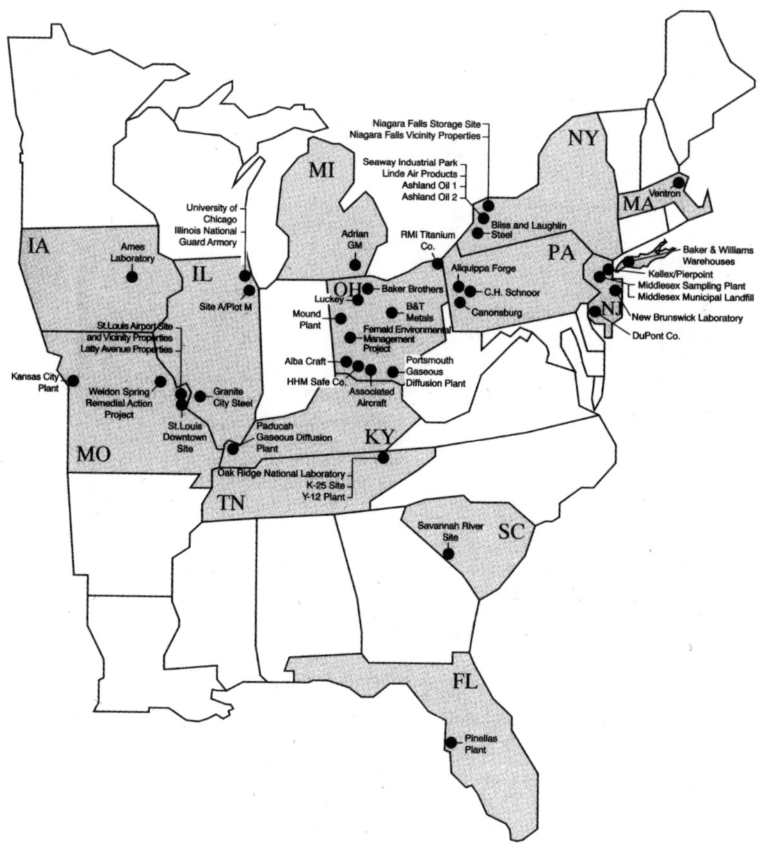

The geologic settings of contaminated sites in the DOE complex are highly variable. DOE installations are located in all major geographic regions of the United States. Table 1-2 shows the geologic and climatologic variability at several of the larger DOE facilities. Site geology, including characteristics of the geologic medium and depth to groundwater, is important for two reasons. First, site geology affects travel times and pathways for contaminant migration in the subsurface. Second, it can be the key factor in determining the performance of a technology designed to clean up subsurface contamination.

TABLE 1-1 Contaminants and Volume of Groundwater, Soil, and Sediment to Be Cleaned up at Major DOE Installations

Installation	Examples of Contaminants of Concern	Estimated Groundwater Volume (m^3)	Estimated Soil and Sediment Volume (m^3)
Savannah River Site	TCE, PCE, aluminum, zinc, arsenic, cadmium, chromium, lithium, mercury, lead, tritium, strontium-90, cesium-137 and 139, cobalt-60	3.1×10^8	8.6×10^6
Hanford Site	Tritium, cobalt, strontium, cesium, technetium, plutonium, uranium, carbon tetrachloride, nitrates, iodine, chromium, mixed waste, transuranic waste	2.0×10^7	6.4×10^7
Oak Ridge Reservation	Asbestos, petroleum hydrocarbons, PCBs, radionuclides (uranium-235 and depleted uranium), mixed waste, strontium-90, cesium-137, cobalt-60, tritium, heavy metals, nitrates, organic solvents, beryllium compounds, mercury, cadmium	4.6×10^6	4.3×10^5
Rocky Flats	Plutonium, americium, uranium, VOCs, PAHs, beryllium (soils); nitrates, metals, solvents (groundwater); radionuclides, metals, VOCs, PCBs (surface water)	1.2×10^6	3.2×10^5
Idaho National Engineering and Environmental Laboratory	Heavy metals, PCBs, acids, asbestos, solvents, low-level radioactive waste, transuranic waste	7.6×10^5	6.5×10^5

NOTE: PAH = polycyclic aromatic hydrocarbon; PCB = polychlorinated biphenyl; PCE = perchloroethylene; TCE = trichloroethylene; VOC = volatile organic compound.

SOURCE: EPA, 1997.

BOX 1-1
Origins of Groundwater and Soil Contamination at DOE Facilities

Groundwater and soil contamination at DOE installations dates from the start of the Manhattan Project to develop nuclear weapons, beginning in 1942. Initially the responsibility of the U.S. Army Corps of Engineers, stewardship of the nation's nuclear arsenal was transferred to the Atomic Energy Commission in 1946 and to the newly created DOE in 1977. DOE inherited responsibility not only for maintaining and increasing the nation's nuclear weapons arsenal, but also for cleaning up the legacy of environmental contamination associated with nuclear weapons production.

The bulk of contamination at DOE installations is a result of nuclear weapons production (DOE, 1997), although some contamination is a by-product of DOE's work on civilian nuclear power projects for the Atoms for Peace program and on nuclear-powered submarines for the Navy. Weapons production processes that ultimately led to groundwater and soil contamination include uranium mining, milling, and refining; isotope separation; fuel and target fabrication; weapons component fabrication; weapons testing; and most important, chemical separations, in which spent nuclear fuel rods and targets were dissolved to extract uranium and plutonium (DOE, 1997).

Many of the contaminants released during the manufacturing of nuclear weapons, including DNAPLs and metals, are similar to those released by major manufacturers of durable goods, such as automobiles and airplanes. However, DOE facilities have the added hazard of radioactive contaminants, which are generally not used in other industries. Like other industries, DOE frequently disposed of wastes in landfills, lagoons, or underground injection wells, and spills of these by-products were not uncommon. These practices ultimately led to widespread groundwater and soil contamination across the weapons complex.

At DOE facilities, the contamination problem was exacerbated by the veil of secrecy and the resultant lack of environmental oversight associated with the nuclear weapons production program (CERE, 1995; DOE, 1997). In part because of policies designed to preserve the secrecy of the nuclear weapons production process, DOE and its predecessor agencies were exempt from environmental laws for most of the nearly five decades between World War II and the end of the Cold War. Dumping of radioactive and hazardous wastes in unauthorized or improperly designed landfills was not uncommon. At Hanford, for example, environmental auditors have discovered unauthorized burial pits—unlined holes dug in the ground—where radioactive wastes were dumped and where no official records of the dumping were maintained (D'Antonio, 1993).

Certain geologic and geochemical characteristics of a site can decrease or increase the migration rates of organic and inorganic contaminants. For example, as described above, DNAPLs can become entrapped in the pore spaces of geologic materials or can sorb (attach) to soils underground, slowing transport. Alternatively, the presence of fractures in the subsurface geologic formation can speed the rate of DNAPL transport. Some geologic formations, such as clayey sand

TABLE 1-2 Geologic and Climatologic Variability Across the DOE Weapons Complex

Installation	Climate	Geology and Hydrogeology	Surface Waters	Depth to Groundwater (m)
Savannah River Site	Humid, subtropical	Atlantic Coastal Plain with clayey soils. The strata are deeply dissected by creeks, and most groundwater eventually seeps into and is diluted by the creeks.	Savannah River	~15
Hanford Site	Arid, cool; mild winters and warm summers; average annual rainfall 16 cm (6.3 in.)	Alluvial plain of bedded sediments with sands and gravels. Groundwater flows toward the Columbia River.	Columbia River	~50-100
Oak Ridge Reservation	Humid, typical of the southern Appalachian region; average annual precipitation 138 cm (54.4 in.)	Valley and Ridge province bordering the Cumberland Plateau. Porosity is low due to fractures. High clay content. Shallow water table.	Clinch River	~2
Rocky Flats	Temperate, semiarid, and continental temperatures; average annual rainfall just under 40 cm (15 in.)	Colorado Piedmont Section of the Plains Physiographic Province. Alluvial deposits cover the installation.	Five streams occur on or near the facility	~15
Idaho National Engineering and Environmental Laboratory	Semiarid with sagebrush-steppe characteristics located in a belt of prevailing western winds; average annual rainfall 22 cm (8.5 in.)	Near the northern margin of the Eastern Snake River Plain, a low-lying area of late Tertiary and Quaternary volcanism and sedimentation. Basalt covers three-quarters of its surface.		~50

SOURCE: Adapted from Sandia National Laboratories, 1996.

with a high sorptive capacity, can slow the mobility of radionuclides. In such cases, the travel time from contaminant release areas to where people may be exposed may be long in comparison to the radionuclide's half-life (tritium, for example, has a half-life of 12.3 years, and strontium-90 has a half-life of 29 years). The geochemistry of the subsurface can also affect contaminant mobility. Under certain geochemical conditions, for example, contaminants dissolving from DNAPLs may biodegrade. Likewise, under some (reducing) conditions, chromium may be present as relatively immobile trivalent chromium, Cr(III), whereas under other (oxidizing) conditions, it may be present as highly toxic and mobile hexavalent chromium. Understanding site geologic and geochemical characteristics is thus critical for predicting contaminant transport, which is a first step in deciding on a cleanup strategy.

Geologic and geochemical characteristics of the site also have a major influence on the performance of subsurface cleanup systems. The subsurface is usually highly heterogeneous, consisting of layers of materials such as sand, gravel, clay, and rock (NRC, 1994). These materials have vastly different abilities to transmit water and other fluids and influence subsurface water chemistry in different ways. Even within a single layer, composition may vary over small distances. Characterizing this variability is extremely difficult because the subsurface cannot be viewed in its entirety (NRC, 1994); hydrogeologic and geochemical properties generally are estimated from samples withdrawn from wells and coring devices placed at discrete intervals. This heterogeneity and difficulty in characterization complicate the design of subsurface cleanup systems because predicting system performance under such uncertain conditions is difficult. Further, many types of cleanup systems, including not only pump-and-treat systems but also systems using in situ chemical oxidation, biodegradation, and other processes, require the circulation of water, aqueous solutions, or other fluids underground. The physical heterogeneity of the subsurface interferes with uniform delivery of fluids to contaminated locations.

Contaminants Found in Groundwater and Soil at DOE Facilities

Understanding the characteristics of contaminants present at DOE installations is also a critical step in determining applicable remediation technologies. Technologies effective for one type of contaminant, such as biodegradable components dissolving from DNAPLs, may not be effective for another type of contaminant, such as a nonbio–degradable radionuclide. Over the past several years, DOE has undertaken numerous studies to characterize groundwater and soil contaminants in the weapons complex (see, for example, DOE, 1992, 1996a; EPA, 1997). However, the nature and extent of groundwater and soil contamination remain poorly understood at many sites.

Tables 1-3 and 1-4 list the most frequently encountered or highest-priority metal, radionuclide, and DNAPL contaminants in groundwater and soil as identified by some of the studies. As is clear from the tables, different studies have reached different conclusions about which

TABLE 1-3 Metals and Radionuclides in Contaminated Groundwater and Soil at DOE Installations

	Source of Information				
				Riley and Zachara, 1992[d]	
Rank	EPA, 1997[a]	SCFA[b]	INEEL, 1997[c]	Metals	Radionuclides
Groundwater					
1	Uranium	Technetium-99	Tritium	Lead	Tritium
2	Tritium	Chromium(VI)	Uranium	Chromium	Uranium-234, 235, 238
3	Thorium	Uranium	Strontium-90	Arsenic	Strontium-90
4	Lead	Tritium	Technetium	Zinc	Plutonium-238, 239, 240
5	Beryllium	Mercury	Chromium	Copper	Cesium-137
6	Plutonium		Cesium-137	Cadmium	Cobalt-60
7	Radium		Beryllium	Barium	Technetium-99
8	Mercury		Lead	Nickel	Iodine-129
9	Arsenic		Thorium	Mercury	
10	Chromium		Plutonium	Cyanide	
Soil					
1	Uranium	Cesium-137	Cesium-137	Copper	Uranium-234, 235, 238
2	Tritium	Uranium	Strontium-90	Chromium	Plutonium-238, 239, 240
3	Thorium	Strontium-90	Uranium	Zinc	Cesium-137
4	Lead	Plutonium	Plutonium	Mercury	Tritium
5	Beryllium	Radium	Cobalt-60	Arsenic	Strontium-90
6	Plutonium	Chromium(VI)	Americium	Cadmium	Thorium-228, 230, 232
7	Radium	Mercury	Tritium	Lead	Cobalt-60
8	Mercury	Thorium	Thorium	Nickel	Technetium-99
9	Arsenic		Lead	Barium	Iodine-129
10	Chromium		Chromium	Cyanide	

[a]The data set includes 86 DOE installations and other locations where characterization and assessment of groundwater and soil have not been completed. The data set does not make separate rankings of contaminants in groundwater and soil.

[b]SCFA developed this ranking based on mobility, prevalence, and toxicity (Jim Wright, Savannah River Site, personal communication, 1997).

[c]This data set is not inclusive across the weapons complex but includes the major waste units identified at about 60 sites in 1995 and 1996. The data were validated in 1997 through review of published references.

[d]The data set includes 91 waste sites at 18 DOE facilities.

TABLE 1-4 Chemicals Present as DNAPLs in Contaminated Groundwater and Soil at DOE Installations

	Source of Information	
Rank	INEEL, 1997[a]	DOE, 1992[b]
Groundwater		
1	Trichloroethylene (TCE)	1,1,1-TCA
2	Dichloroethylene (DCE)	1,2-DCE
3	Perchloroethylene (PCE)	PCE
4	Vinyl chloride	1,1-DCA
5	Trichloroethane (TCA)	Chloroform
6	Chloroform	1,1-DCE
7	Dichloroethane	Carbon tetrachloride
8	Carbon tetrachloride	1,2-Dichloromethane
Soil		
1	TCE	TCE
2	Polychlorinated biphenyls	1,1,1-TCA
3	DCE	PCE
4	PCE	Dichloromethane
5		Carbon tetrachloride
6		Chloroform
7		Freon
8		1,2-DCA
9		1,1,2,2-Tetrachloroethane
10		Chlorobenzene

[a] This data set is not inclusive across the weapons complex but includes the major waste units identified at about 60 sites in 1995 and 1996. The data were validated in 1997 through review of published references.

[b] The data set includes 91 waste sites at 18 DOE facilities.

contaminants are most prevalent or highest priority. Nevertheless, all studies indicate that uranium, technetium, strontium-90, and tritium are commonly detected radionuclides; chromium is generally the key metal contaminant; and trichloroethylene (TCE) and other solvents are commonly encountered DNAPLs. A precise ranking of key contaminants of concern, based on either risk or prevalence, is not possible, given the limitations of existing data.

Complicating the design of treatment systems for many DOE sites is the presence of mixtures of contaminants. In a survey of 91 DOE waste sites, for example, Riley and Zachara (1992) found that mixtures of two or more compounds were present at 59 (65 percent) of the sites. In soils, the most frequently occurring mixtures were metals combined with radionuclides, but various combinations of metals and radionuclides with organic contaminants were also observed at some sites. In groundwater, the most common mixtures were metals and chlorinated hydrocarbons.

Risks of Groundwater and Soil Contamination at DOE Installations

Risks to human health and the environment are the basis for laws requiring the remediation of subsurface contaminants at DOE installations. Remediation technologies must be designed to reduce these risks. Risk, in addition to geologic and contaminant characteristics, is thus the third dimension of the subsurface contamination problem that must be understood in order to determine which types of new remediation technologies are needed most. Quantitative information on health and ecological risks of groundwater and soil contamination at DOE installations is limited.

Conduct of comprehensive risk assessments at DOE installations is complicated by the difficulty in characterizing contaminant transport pathways and doses potentially received by people and ecosystems near the installation. Exposure to groundwater and soil contaminants from DOE installations might occur through multiple possible pathways (see, for example, Figure 1-3). Characterizing the locations and concentrations of subsurface contaminants and their transport along pathways to potential receptors has proved to be a daunting task.

In general, quantitative assessments of the full risks posed by each installation's contamination have not been conducted, with the exception of studies at Fernald and, to a lesser extent, Oak Ridge (CERE, 1995). Available quantitative risk information is generally limited to studies of discrete contaminated sites within each installation, usually conducted as part of the cleanup process under CERCLA. The aggregate nature of risks from each facility, given all the sites within a facility, is unknown (CERE, 1995).

The most comprehensive study to date of risks posed by contamination at DOE facilities was conducted by the Consortium for Environmental Risk Evaluation (CERE), organized by Tulane and Xavier Universities at DOE's request. This study, completed in 1995, was part of a broader DOE effort to prioritize site cleanup activities according to health and ecological risks. CERE reviewed existing health and ecological risk studies at six major DOE installations: Hanford, Idaho National Engineering and Environmental Laboratory, Oak Ridge, Rocky Flats, the Savannah River Site, and Fernald. Boxes 1-2 and 1-3 summarize some of the groundwater and soil contamination risks at these installations, according to CERE's study.

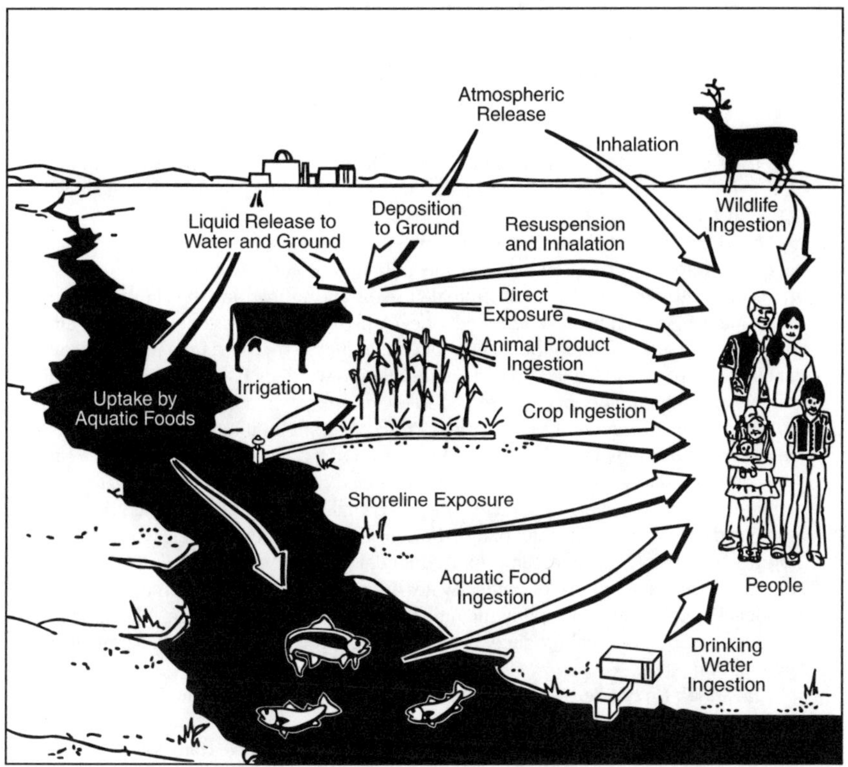

FIGURE 1-3 How people could have been exposed to radioactive materials from Hanford. SOURCE: Technical Strategies Panel, Hanford Environmental Dose Reconstruction Project, 1994.

CERE was unable to quantify the magnitude of the health risks posed by DOE installations. CERE concluded that groundwater and soil contamination appear to present little or no immediate hazard to most populations neighboring DOE installations. However, Native Americans living near Hanford are currently at risk, and there is potential for significant future risks if the contamination is left uncontrolled and restrictions on facility access are lifted. CERE's report advises, "Without careful management, there could be significant risks to workers, to the public and nearby tribes . . . from plutonium, spent nuclear fuel, and nuclear wastes currently stored at DOE installations" (CERE, 1995).

Just as the CERE study was unable to quantify the magnitude of the human health risks posed by the weapons complex as a whole, it was unable to place quantitative bounds on the level of ecological

BOX 1-2
Health Risks of Subsurface Contamination at Select DOE Installations

The Consortium for Environmental Risk Evaluation, in a 1995 study, reported the following information about health risks due to subsurface contamination at select DOE installations.

Hanford Site. At Hanford, there is a potential risk to Native Americans who are allowed, through treaty rights, to use the Columbia River and its banks for subsistence purposes and to access contaminated seeps and springs and the Hanford town site. Native Americans have reportedly developed rashes consistent with exposure to chromium after being in the vicinity of contaminated seeps. If contaminated soil at Hanford were not cleaned up and the installation were open to unrestricted use in the future, risks to other occasional and frequent users would be potentially high.

Fernald Environmental Management Project. Some uranium from Fernald has migrated off property in groundwater to the south of the installation. Residents near the south boundary who use private wells are being provided with bottled water. In addition, uranium levels in surface soils on Fernald property can be hundreds to thousands of times greater than natural background levels; these levels would translate to high risks if people came into continual contact with the soils. For now, access restrictions and use of bottled water appear to be controlling risks, although models indicate that off-property migration of contaminants may pose risks through the consumption of local produce, beef, and milk grown or raised on contaminated agricultural land.

Savannah River Site. Most areas at the Savannah River Site having high contaminant concentrations are several miles or more from installation boundaries. The potential for off-site contaminant transport is low due to the large transport distances involved, the high levels of dilution, and the generally low mobility in groundwater and soils of many of the major contaminants. No significant health risks to the general public appear to exist under current conditions. Although the DOE has identified several large plumes of contaminated groundwater at the installation, current and planned remedial actions have been designed to control off-site migration of the plumes at the two known areas where such migration appears possible in the near future.

Rocky Flats Environmental Technology Site. Because of its proximity to a major population center (Denver), Rocky Flats operates an aggressive installation-wide monitoring program, and monitoring has shown that there are no current off-site exposure risks to the public from groundwater contamination. However, predictions indicate that contaminated groundwater will migrate off-site in 30 to 300 years if no action is taken to control contamination. Similarly, on-site soils currently pose no risks to the public because of access restrictions. However, if unrestricted access occurs prior to remediation, potential risks would be significant, due to high levels of radionuclides in the soil at portions of the installation.

Oak Ridge Reservation. Under current conditions at Oak Ridge, exposure to contaminants in groundwater could occur either via groundwater discharge to the Clinch River or via direct off-site migration of the groundwater, which has occurred in Union Valley. However, according to CERE, radioactive and nonradioactive contaminants in the Clinch River occur at concentrations that "pose a low risk from exposure either

continues on next page

in drinking water or in contact through recreational use." Contaminant levels in groundwater plumes that have migrated off-site are below drinking water standards. Risks could occur in the future if groundwater contaminants are allowed to migrate off-site at high concentrations or if restrictions on land use are lifted.

Idaho National Engineering and Environmental Laboratory (INEEL). INEEL is isolated, so the potential for public exposure to site contaminants is limited, and strong institutional controls are currently in place to limit access. The primary possible route of public exposure to contamination in the future is via migration of contaminants to the Snake River Plain Aquifer. However, CERE concluded that the risk of such exposure is low because of the long distances between INEEL and the nearest water supply wells and natural processes (such as biodegradation, radioactive decay, and dilution) that decrease contaminant concentrations along potential migration pathways.

SOURCE: Summarized from CERE, 1995.

BOX 1-3
Ecological Risks of Subsurface Contamination at Select DOE Installations

The Consortium for Environmental Risk Evaluation, in its 1995 study, reported the following information about ecological risks due to subsurface contamination at select DOE installations.

Hanford Site. According to the limited available ecological risk information, risks due to contaminated soils at Hanford are primarily in the vicinity of major operations at the installation. Ecosystems in these areas were initially disturbed during construction of Hanford's facilities. The primary ecological risks of concern due to groundwater contamination at Hanford are possible effects on salmon spawning areas from the discharge of contaminated groundwater into the Columbia River. Concentrations of contaminants in spawning areas could in theory approach those in groundwater because several areas are near points at which plumes of contamination emanating from former reactors discharge to the Columbia River. For example, concentrations of chromium(VI) in some plumes discharging to the river are more than 25 times the concentration known to damage juvenile salmon. So far, however, the magnitude of this risk has not been determined.

Fernald Environmental Management Project. Ecological risk studies conducted after completion of the CERE study did not identify significant ecological risks due to groundwater and soil contamination at Fernald. Onsite, metal contaminants are found in all media, but only uranium and molybdenum have been detected at levels above the benchmark toxicity value for ecological risk as reported in ecological literature. Offsite, only uranium has been detected at above benchmark ecological toxicity values, and the highest levels of uranium are less than twice the benchmark toxicity value. No excess radiation has been detected in plants or animals on- or offsite.

continues on next page

Savannah River Site. Ecological risks are better characterized at the Savannah River Site than at any other DOE installation, due in part to the designation of the site as a national environmental research park and the presence of the Savannah River Ecology Laboratory. Researchers at the installation have detected no significant risks offsite due to groundwater and soil contamination, though they have indicated that more information is needed to understand the migration of nonradioactive contaminants into the Savannah River Swamp, transport of contaminants offsite with biota, and effects of soils contaminated with solvents. The influence of groundwater seeps on surface water at the installation also has not been well studied. One episode of plants' dying due to exposure to contaminated groundwater from a seep has been reported. Future land use decisions are key in judging the level of ecological risk. The CERE study concluded that "human encroachment of the Savannah River Site is the source of greatest ecological risk to the ecosystem."

Rocky Flats Environmental Technology Site. Ecological risk studies at Rocky Flats were quite limited when CERE conducted its study. Based on the limited studies available, CERE was unable to identify significant ecological risks due to groundwater or soil contamination. CERE concluded that the major ecological risk to the site, which is home to seven endangered and threatened species, is industrialization.

Oak Ridge Reservation. Although extensive ecological studies have been conducted at Oak Ridge, limited information is available on risks due to groundwater and soil contamination. The existing studies focused on surface water systems, bottomland hardwoods, and "old field" communities (ecosystems that have evolved on previously cleared land). Some accumulation of mercury and polychlorinated biphenyls has been observed at top levels of the terrestrial food chain, perhaps due in part to uptake of contaminants by plants.

Idaho National Engineering and Environmental Laboratory. Groundwater transport of contaminants at Idaho is very slow and, CERE concluded, probably does not have a significant impact on ecosystems. The major off-site transport routes for contamination are wind-blown dusts and waterfowl migration, but studies carried out to date have concluded that concentrations are too low for ecological effects to occur as a result of these migration pathways. As at other installations, however, large data gaps and uncertainties make these conclusions tentative.

SOURCE: Summarized from CERE, 1995.

risk posed by groundwater and soil contamination (see Box 1-3). However, CERE concluded that although contamination of biota, soil, sediment, and water resources is widespread across the weapons complex, ecological effects generally appear to be confined to localized contaminated areas. Ecosystems in these areas may be as much or more affected by the original construction and manufacturing activities that took place at the various installations as by the presence of contamination, CERE suggests. Further, some planned remediation activities may cause additional harm to ecosystems. For example, soil cleanup plans at Fernald call for digging up surface soil across nearly 50 percent of the installation, which will destroy the native ecosystem in excavated areas. At Oak Ridge and elsewhere, plans for capping some waste areas and for in situ vitrification of others will limit the ability of these areas to support plant and animal life. Some DOE installations, because of restrictions placed on public access, have unique ecosystems that have been preserved virtually undamaged; several also support endangered and threatened plant and animal species. For example, seven threatened and endangered species, including the bald eagle and peregrine falcon, have been sighted at Rocky Flats. Some potential remedies for these sites, as well as the possible future lifting of land use restrictions, could jeopardize important habitats. A major limitation of CERE's study is the lack of sufficient understanding of the ecosystems (for example, lack of data showing changes over time) at most DOE installations. Thus, additional studies are needed before definitive conclusions can be reached about ecological risks due to groundwater and soil contamination in the weapons complex.

Around the time that CERE completed its study, DOE undertook a pilot-scale project to develop improved methods for quantifying human health risks across the weapons complex (Hamilton, 1994, 1995). The pilot project focused on a few contamination problems at the Savannah River Site, Fernald, and the Nevada Test Site. At Fernald and the Nevada Test Site, investigators calculated potential health risks associated with drinking contaminated groundwater. In both instances, off-site risks were within or below EPA's acceptable range of excess individual lifetime cancer risk (10^{-4} to 10^{-6}). At Fernald, the highest predicted lifetime risk for an individual consuming well water was 1.3×10^{-5} for a well placed near the installation boundary. At the Nevada Test Site, the lifetime risk of cancer mortality due to exposure to radionuclides in groundwater for persons living off-site was projected at 7×10^{-7}. However, the risk from consuming well water on site was 7×10^{-3}, pointing to the importance of maintaining restrictions on site access.

DOE'S PROGRESS TO DATE IN CLEANING UP GROUNDWATER AND SOIL CONTAMINATION

Estimating DOE's progress in cleaning up contaminated groundwater and soil is difficult because of conflicting terminology. DOE's Office of Environmental Restoration generally tracks groundwater and soil cleanup projects by "operable unit" or "project"—that is, a subset of an installation at which contaminated groundwater and/or soil problems are being cleaned up as one unit. Each installation generally has a number of operable units or projects. However, the Office of Environmental Management reports information on groundwater and soil contamination in terms of "release sites"—that is, the number of individual sites at which contaminants were released. One operable unit may encompass more than one release site. Thus, data on the number of operable units at which cleanup plans are under way cannot be compared directly to data on the total number of individual sites at which contaminant releases have been reported.

DOE has estimated that a total of 10,000 sites must be cleaned up at its installations (EPA, 1997). Cleanup plans are being prepared for groundwater contamination at a total of 92 projects (or operable units), according to recent DOE data (Tolbert-Smith, 1998). Remedies have been selected for 27 (29 percent) of these projects (Tolbert-Smith, 1998). Similarly, soil cleanup plans are under way for 221 projects, and remedies have been selected for 163 (74 percent) of these projects (Tolbert-Smith, 1998). It is unclear what percentage of the 10,000 contaminated sites is being addressed by these remedies. Based on reports to the committee and visits to DOE installations, it appears that most of the groundwater and soil remediation work remains to be completed.

DOE's recent budget projections have assumed that most groundwater will not be cleaned up, in part because of technical limitations. DOE's 1996 budget assessment, presented in the *1996 Baseline Environmental Management Report* (DOE, 1996a), assumed that sources of groundwater contamination would be removed and pump-and-treat technologies would be used where effective, but that otherwise contaminated groundwater would simply be contained on-site and monitored, rather than cleaned. For example, the budget projection assumed that groundwater contamination at Hanford and the Idaho National Engineering and Environmental Laboratory would be managed by a combination of limited pumping and treating followed by monitoring of remaining contamination. The 1998 budget assessment, entitled *Accelerating Cleanup: Paths to Closure*, assumes that groundwater remediation will be considered complete when the contamination is contained or when a long-term treatment or monitoring system is in place (DOE, 1998).

Although DOE's budget projections have discounted the problems of contaminated groundwater, it is still obliged to meet the requirements of federal statutes requiring groundwater cleanup. Further, political pressure to clean up these resources will remain. Recently,

DOE has faced considerable pressure from members of Congress to shore up its efforts to protect groundwater from contamination by radioactive wastes leaking from underground storage tanks, for example (GAO, 1998). Thus, DOE cannot avoid its groundwater and soil contamination problems or the limitations in technologies for addressing them.

REFERENCES

Betts, K. S. 1998. Energy efficiency research gains in Department of Energy's 1999 budget request. Environmental Science & Technology 2(4):167A.

CBO (Congressional Budget Office). 1994. The Total Costs of Cleaning up Nonfederal Superfund Sites. Washington, D.C.: U.S. Government Printing Office.

CERE (Consortium for Environmental Risk Evaluation). 1995. Health and Ecological Risks at the U.S. Department of Energy's Nuclear Weapons Complex: A Qualitative Evaluation. New Orleans: Tulane University Medical Center.

D'Antonio, M. 1993. Atomic Harvest: Hanford and the Lethal Toll of America's Nuclear Arsenal. New York: Crown Publishers, Inc.

DOE (Department of Energy). 1992. Chemical Contaminants on DOE Lands and Selection of Contaminated Mixtures for Subsurface Science Research. DOE/ER-0547T. Washington, D.C: DOE, Office of Energy Research.

DOE. 1996a. The 1996 Baseline Environmental Management Report. Washington, D.C.: DOE, Office of Environmental Management.

DOE. 1996b. Subsurface Contaminants Focus Area Technology Summary. DOE/EM-0296. Washington, D.C.: DOE, Office of Science and Technology.

DOE. 1997. Linking Legacies: Connecting the Cold War Nuclear Weapons Processes to Their Environmental Consequences. DOE/EM-0319. Washington, D.C.: DOE.

DOE. 1998. Accelerating Cleanup: Paths to Closure. Washington, D.C.: DOE, Office of Environmental Management.

EPA (Environmental Protection Agency). 1997. Cleaning up the Nation's Waste Sites: Markets and Technology Trends. EPA 542-R-96-005. Washington, D.C.: EPA, Office of Solid Waste and Emergency Response.

EPA. 1998. Treatment Technologies for Site Cleanup: Annual Status Report (Ninth Edition). EPA-542-R98-018. Number 9. Washington, D.C.: EPA, Office of Solid Waste and Emergency Response.

GAO (U.S. General Accounting Office). 1997. Department of Energy: Funding and Workforce Reduced, but Spending Remains Stable. GAO/RCED-97-96. Washington, D.C.: GAO.

GAO. 1998. Nuclear Waste: Understanding of Waste Migration at Hanford Is Inadequate for Key Decisions. GAO/RCED-98-80. Washington, D.C.: GAO.

Hamilton, L. D. 1994. Pilot study risk assessment for selected problems at three U.S. Department of Energy facilities. Environment International 20(5):585-604.

Hamilton, L. D. 1995. Lessons learned: Needs for improving human health risk assessment at USDOE sites. Technology Journal of the Franklin Institute 332: 15-33.

Harden, B. 1996. Half-lives: Government, poison, lies and the splitting of atomic America. Washington Post Magazine (May 5):12-19; 26-29.

INEEL (Idaho National Engineering and Environmental Laboratory). 1997. Decision Analysis for Remediation Technologies (DART) Data Base and User's Manual. INEEL/EXT-97-01052. Idaho Falls: INEEL.

MacDonald, J. A., and M. C. Kavanaugh. 1994. Restoring contaminated groundwater: An achievable goal? Environmental Science and Technology. 28(8):362A-368A.

NRC (National Research Council). 1994. Alternatives for Ground Water Cleanup. Washington, D.C.: National Academy Press.

NRC. 1997. Innovations in Ground Water and Soil Cleanup: From Concept to Commercialization. Washington, D.C.: National Academy Press.

Probst, K. N., and M. H. McGovern. 1998. Long-Term Stewardship and the Nuclear Weapons Complex: The Challenge Ahead. Washington, D.C.: Resources for the Future.

Riley, R. G., and J. M. Zachara. 1992. Chemical Contaminants on DOE Lands and Selection of Contaminant Mixtures for Subsurface Science Research. DOE/ER-0547T. Washington, D.C.: Department of Energy, Office of Energy Research.

Sandia National Laboratories. 1996. Performance Evaluation of the Technical Capabilities of DOE Sites for Disposal of Mixed Low-Level Waste. SAND96-0721/1 UC-2020. Albuquerque, N.M.: Sandia National Laboratories.

Technical Strategies Panel, Hanford Environmental Dose Reconstruction Project. 1994. Summary: Radiation Dose Estimates from Hanford Radioactive Material Releases to the Air and the Columbia River. Atlanta, Ga.: Centers for Disease Control.

Tolbert-Smith, L. 1998. Unpublished data on groundwater and soil contamination remedies at DOE installations. Washington, D.C.: U.S. Department of Energy.

2

The Changing Regulatory Environment

The driver for the Department of Energy's (DOE's) groundwater and soil cleanup technology development program is the need to meet applicable federal and state regulations for the cleanup of contaminated sites. DOE will not be able to meet all of the applicable regulations with existing remediation technologies and therefore must develop new technologies, as discussed in Chapter 1. An understanding of the cleanup goals required under existing regulations is thus critical to the administration of the Subsurface Contaminants Focus Area (SCFA) because the technologies developed by SCFA will have to be capable of meeting applicable regulatory requirements.

Historically, regulators have used drinking water standards as baseline cleanup goals for contaminated groundwater. For soil, commonly regulators have set cleanup goals designed to protect the groundwater beneath the soil and to prevent exposure to contamination via soil ingestion or inhalation. However, these policies are changing rapidly. New policies for groundwater and soil cleanup will affect the range of remediation technology options that DOE can consider using at its sites and will therefore influence SCFA's priorities for technology development. This chapter first describes regulations that prescribe how groundwater and soil are to be cleaned up at DOE installations. It then reviews baseline cleanup goals for groundwater and soil contaminants under these regulations and describes in detail new policies that allow alterations to the historical cleanup process.

OVERVIEW OF APPLICABLE FEDERAL REGULATIONS

During most of the period of nuclear weapons production, operations at DOE installations were managed under a veil of secrecy. DOE remained exempt from federal, state, and local solid and hazardous waste laws that were developed and applied to private industries. The initial spur for DOE to act on its legacy of environmental contamination came largely from two successful lawsuits against the agency, one filed by the Legal Environmental Assistance Foundation and decided in 1984, and another filed by the Natural Resources Defense Council and decided in 1989 (Probst and McGovern, 1998). These suits required DOE to comply with requirements of the Resource Conservation and Recovery Act, which governs management of contaminants from active waste treatment, storage, or disposal facilities. However, DOE remained exempt from other solid and hazardous waste laws until 1992, when Congress passed the Federal Facilities Compliance Act. Cleanup of DOE's installations is therefore still a relatively new undertaking.

Table 2-1 lists laws applicable to the cleanup of groundwater and soil contamination at DOE installations. The listing is in order of significance to DOE remediation projects, rather than in chronological order.

The Federal Facilities Compliance Act, listed first in Table 2-1, is the key law underlying all DOE cleanups because, as mentioned above, it established for the first time that DOE must comply with existing environmental statutes. The act makes federally owned and operated facilities subject to state-imposed fines and penalties for violation of hazardous waste requirements. It authorizes environmental regulators to treat DOE facilities like privately owned industrial facilities and to subject them to the same rules and liabilities.

For cleanup of contaminated groundwater and soil at DOE installations, the most important laws are the Resource Conservation and Recovery Act (RCRA) and the Comprehensive Environmental Response, Compensation and Liability Act (CERCLA), listed second and third in Table 2-1. RCRA, enacted in 1976 and significantly amended in 1984, addresses the treatment, storage, and disposal of hazardous waste at operating facilities. The 1984 amendments brought "solid waste management units," which are currently inactive but formerly used hazardous waste disposal sites within the boundary of an operating facility, under the umbrella of RCRA, as well. CERCLA, enacted in 1980 and amended in 1986, governs the cleanup of groundwater and soil at inactive facilities. Most DOE installations are currently regulated under CERCLA, RCRA, or both (see Table 2-2). Individual

TABLE 2-1 Federal Regulations Applicable to DOE Remediation Projects

Federal Law (year)	General Description	Significance to DOE Remediation
Federal Facilities Compliance Act (1992)	Establishes a definition of "mixed waste" and requires development of plans for its management. Waives the DOE (and other federal facilities') immunity from EPA and state hazardous waste regulations and sanctions.	Establishes that DOE facilities are subject to and liable under federal and state waste management regulations, including CERCLA and RCRA.
Resource Conservation and Recovery Act (1976); Hazardous and Solid Waste Amendments (1984)	Establishes regulations controlling generation, transportation, treatment, storage, and disposal of hazardous materials for active industrial facilities.	Contains requirements for groundwater monitoring. If groundwater contamination is identified, the RCRA corrective action process is implemented.
Comprehensive Environmental Response, Compensation, and Liability Act (1980); Superfund Amendment and Reauthorization Act (1986)	Establishes regulations for cleanup of inactive hazardous waste sites and determines the distribution of cleanup costs among the parties that generated and handled the hazardous substances disposed at these sites.	Requires cleanup of inactive facilities and establishes procedures and requirements for site characterization and remedy selection. SARA provides additional requirements for documentation and distribution of information on releases of pollutants from facilities.
Uranium Mill Tailings Remediation Control Act (1978), and amendments	Establishes two programs to protect public health and the environment from exposure to contaminants from uranium mill tailings piles: one program addresses inactive sites; the other addresses active sites that are licensed by the Nuclear Regulatory Commission.	Requires that DOE take charge of cleaning up contaminated groundwater and soil at 24 inactive uranium mining and processing sites.

(continues on next page)

TABLE 2-1 Continued

Federal Law (year)	General Description	Significance to DOE Remediation
Safe Drinking Water Act (1974)	Develops drinking water standards that govern the quality of water delivered to the consumer and establishes the underground injection control program, which classifies and regulates types of injection well practices.	If groundwater beneath a DOE facility can be used as drinking water, then cleanup levels must satisfy MCLs for chemicals regulated under the SDWA.
Clean Water Act (1972), and amendments	Establishes requirements controlling discharge of pollutants to surface waters.	If groundwater beneath a DOE facility discharges to surface water bodies, then the CWA may be used to establish cleanup levels.
Toxic Substances Control Act (1976), and amendments	Establishes responsibility of manufacturers to provide data on health and environmental effects of chemical substances and provides EPA with authority to regulate manufacture, use, distribution, and disposal of chemical substances.	Includes special management provisions for handling and cleaning up material containing PCBs, which are present in environmental media at DOE facilities.

NOTE: CWA= Clean Water Act; EPA = Environmental Protection Agency; MCL = maximum contaminant level; PCB = polychlorinated biphenyl; SARA = Superfund Amendment and Reauthorization Act; SDWA = Safe Drinking Water Act

TABLE 2-2 Primary Regulatory Drivers for Groundwater and Soil Remediation at Select DOE Installations

Installation	Primary Regulatory Driver: CERCLA	Primary Regulatory Driver: RCRA
Fernald	x	
Hanford	x	x
Idaho National Engineering and Environmental Laboratory	x	x
Lawrence Livermore National Laboratory	x	
Los Alamos National Laboratory		x
Mound Plant	x	x
Nevada Test Site		x
Oak Ridge Reservation	x	x
Paducah Gaseous Diffusion Plant	x	x
Pantex Plant	x	x
Rocky Flats Plant	x	x
Sandia National Laboratories		x
Savannah River Site	x	x

SOURCE: DOE. 1998. Remediation Action Program Information Center (RAPIC). Oak Ridge, Tenn.: Oak Ridge National Laboratory, RAPIC (http://www.em/doe.gov/rapic).

cleanup sites may be regulated under both programs in situations where cleanup began under RCRA but where the installation was later listed on the National Priorities List (GAO, 1994). In other cases, RCRA cleanup sites may be very near, even next to, sites where cleanup is occurring under CERCLA (GAO, 1994). For example, at Hanford, the B-pond disposal site for liquid wastes is being cleaned up under RCRA, while abandoned trenches that once carried wastes to the pond are being cleaned up under CERCLA (GAO, 1994).

RCRA established a manifest program to track hazardous waste at active facilities from the point of generation through transport to treatment, storage, and disposal. The program is administered through a system of permits. To obtain a permit to operate a facility that is subject to RCRA, the facility owner must monitor the groundwater beneath and downgradient of the operation to determine if statistically significant increases in contaminant concentrations (higher than those occurring upgradient of the site) exist. If the monitoring program indicates that contaminant concentrations are increasing, the facility owner must determine whether concentrations exist at levels above predetermined groundwater protection standards. Where such standards are exceeded, the site owner must implement a "corrective

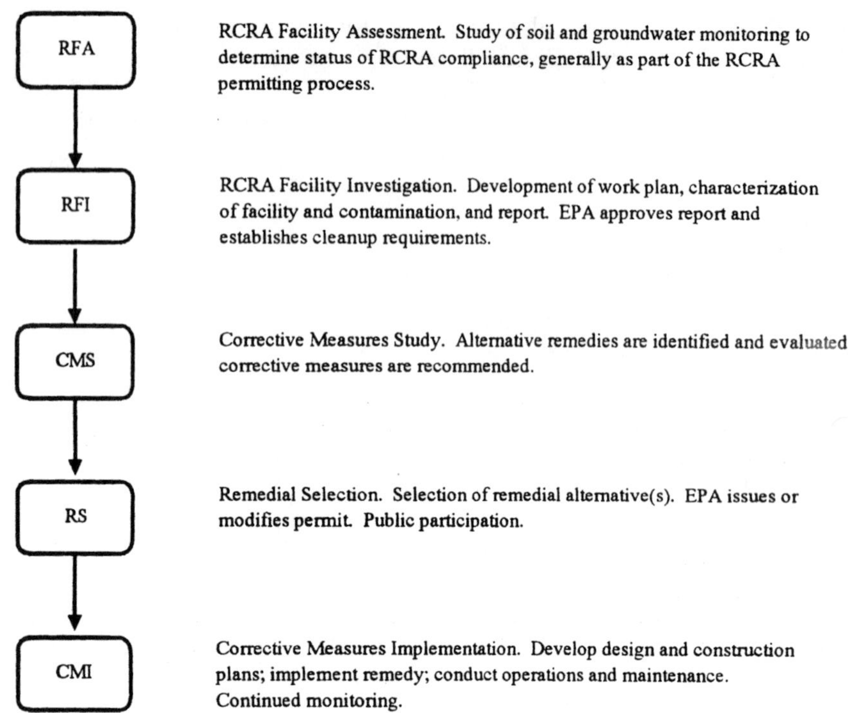

FIGURE 2-1 Steps in the RCRA corrective action process. NOTE: EPA = Environmental Protection Agency.

action" program to clean up the contaminated groundwater and soil. Figure 2-1 shows the steps in the RCRA corrective action process.

CERCLA authorized the federal government to require cleanup of abandoned or inactive facilities where groundwater and soil are contaminated. Under CERCLA, current and former site owners can be held liable for cleanup costs. CERCLA also established a federal fund, Superfund, to pay for cleanup of sites where responsible parties cannot be identified. Because CERCLA facilities are no longer active, the program, unlike RCRA, is not operated through a permit system. Rather, the federal government is charged with identifying the nation's most highly contaminated sites, listing them in a data base known as the National Priorities List (NPL), and ensuring that the sites are cleaned up. Figure 2-2 outlines the CERCLA remedial process.

Although the regulatory mechanisms under RCRA and CERCLA differ, the processes for selecting cleanup remedies under the two

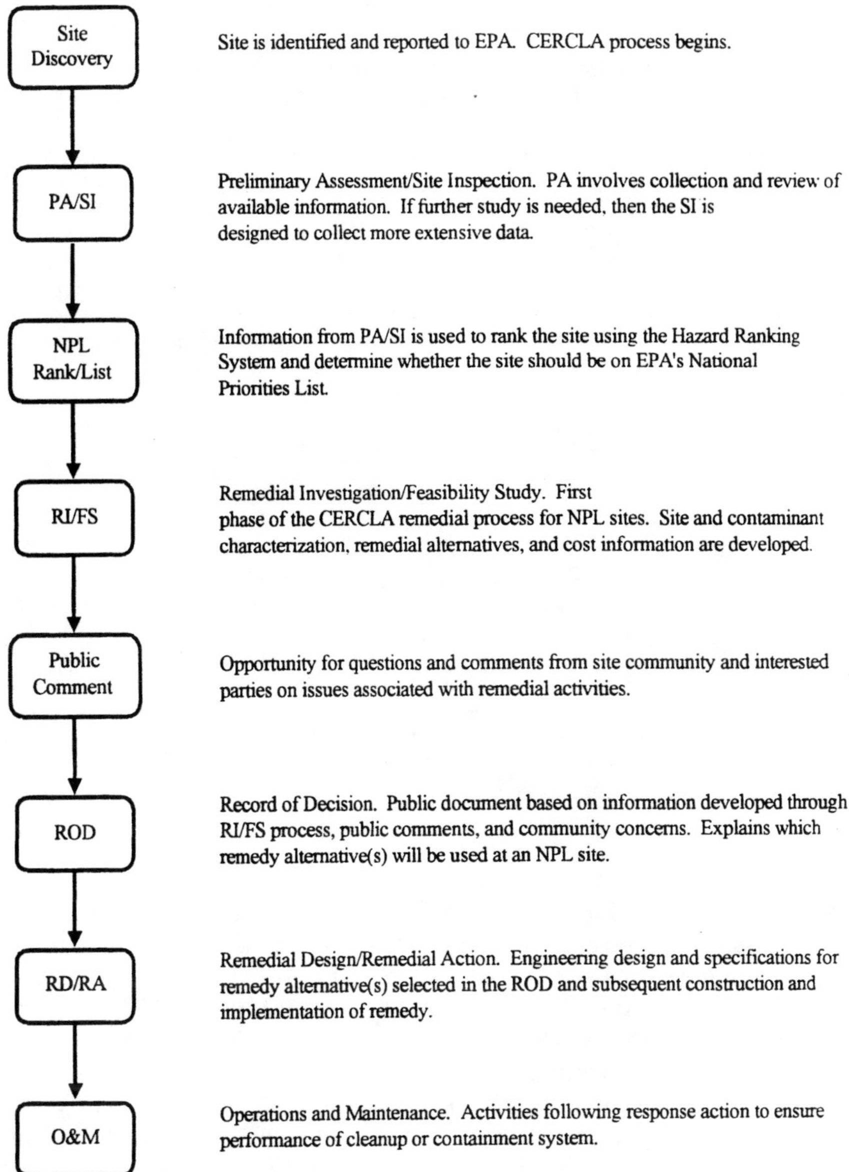

FIGURE 2-2 Steps in the CERCLA remedial process. NOTE: EPA = Environmental Protection Agency.

BOX 2-1
Feasibility Study Criteria for Evaluation of Alternative Remedies Under CERCLA

CERCLA regulations require consideration of the nine evaluation criteria listed below when selecting among possible cleanup remedies. Not all of these criteria receive equal weight. The first two are threshold requirements. Any remedy selected must be protective of human health and the environment, and "applicable or relevant and appropriate requirements" (ARARs—other federal, state, or tribal laws that apply to a particular site cleanup) must be followed. The next five criteria are considered balancing criteria. The selected remedy must be cost effective and use permanent solutions and treatment to the maximum extent practicable. The last two criteria are modifying criteria.

Threshold Criteria
- Overall protection of human health and environment
- Compliance with ARARs

Balancing Criteria
- Long-term effectiveness and permanence
- Reduction of toxicity, mobility, or volume
- Short-term effectiveness
- Implementability
- Cost effectiveness

Modifying Criteria
- Regulatory acceptance
- Community acceptance

programs are similar. For example, CERCLA's remedial investigation/feasibility study corresponds to the RCRA facility investigation/corrective measures study (see Figures 2-1 and 2-2). Development of the CERCLA record of decision (ROD) corresponds to the RCRA remedy selection step. The two programs also require consideration of similar criteria when selecting cleanup remedies. CERCLA requires consideration of nine evaluation criteria (see Box 2-1), and nearly the same set of criteria are used in remedy selection under RCRA. The groundwater and soil cleanup goals under the two programs are similar as well, as discussed below. A significant difference between the two programs is CERCLA's inclusion of a specific step in the time sequence of remedy selection for public comment; opportunities for public comment are less prominent under RCRA.

The fourth law listed in Table 2-1, the Uranium Mill Tailings Remediation Control Act (UMTRCA), applies to 24 former uranium ore mining and processing sites where cleanup is being overseen by DOE (EPA, 1995). These sites generally consist of mine tailings piles and groundwater and soil originating from these piles. When Congress enacted this legislation in 1978, members of Congress directed DOE to take charge of cleaning up 22 sites specifically identified in the legislation; DOE has since added two more sites to the list. The Environmental Protection Agency (EPA) oversees DOE's cleanup of UMTRCA sites, in cooperation with the Nuclear Regulatory Commission, according to a cleanup process and set of cleanup standards developed specifically for UMTRCA . Because UMTRCA applies to a small and unique subset of DOE sites, most of the discussion in this chapter focuses on RCRA and CERCLA.

The Safe Drinking Water Act (SDWA) and the Clean Water Act (CWA), listed fifth and sixth in Table 2-1, do not directly regulate cleanup of contaminated groundwater and soil at DOE installations but rather provide the basis for setting groundwater and soil cleanup goals. The Toxic Substances Control Act, listed last in the table, has limited applicability to DOE groundwater and soil remediation projects; it includes special management provisions for cleanup of material containing polychlorinated biphenyls (PCBs), found as environmental contaminants at many DOE facilities.

BASELINE CLEANUP GOALS

Prior to assessing which remedial technologies might be applicable to a site, site managers must determine the remediation goals, including the targeted cleanup concentrations for each constituent in groundwater, soils, and other media, as appropriate. The baseline standards used as cleanup goals vary somewhat depending on the regulatory program and the individual regulators involved in overseeing a specific site, but in general, cleanup goals under CERCLA and RCRA are intended to be consistent.

Groundwater Baseline Cleanup Goals

Guided by the Safe Drinking Water Act, EPA and state agencies have established public drinking water standards for many compounds. EPA refers to these standards as "maximum contaminant levels" (MCLs). MCLs are established based on risk (toxicity and carcinogenicity), capability of drinking water treatment technologies to remove the particular contaminant, and cost considerations. MCLs apply to drinking water supplies.

Groundwater may serve as a drinking water supply; therefore

TABLE 2-3 Maximum Contaminant Levels for Significant DOE Contaminants

Contaminant	MCL
Constituents from DNAPLs	
Trichloroethylene	0.005 mg/liter
Tetrachloroethylene	0.005 mg/liter
Vinyl chloride	0.002 mg/liter
cis-1,2-Dichloroethylene	0.07 mg/liter
Metals	
Chromium (total)	0.1 mg/liter
Uranium	0.020 mg/liter
Radioactivity	
Beta particle and photon activity	4 mrem
Gross alpha particle activity	15 pCi/liter
Combined Ra-226 and Ra-228	5 pCi/liter
Radon (1991 proposed)	300 pCi/liter

the most common cleanup standards for groundwater at RCRA and CERCLA sites are MCLs. EPA has developed MCLs for organic chemicals, inorganic chemicals, and radionuclides. Table 2-3 lists the MCLs for organic chemicals that can dissolve from dense nonaqueous-phase liquids (DNAPLs) frequently identified at DOE facilities and for chromium, uranium, and radioactivity measures. The MCL for uranium is based on toxicity rather than radioactivity. For radionuclides, such as strontium-90, that emit beta particles and gamma rays, the MCL is based on dose equivalent (the effective radiation dose that the body receives), whereas for other radionuclides, the MCL is based on activity (the number of disintegrations of the radioactive compound per unit time).

Within the RCRA program, provisions have been made for development of site-specific "alternate concentration limits" (ACLs) that can serve as cleanup goals for contaminated groundwater in place of MCLs. A corresponding procedure is available under CERCLA as part of the process of evaluating what are known as "applicable or relevant and appropriate requirements" (ARARs, which are environmental laws other than CERCLA that might apply to the site and that must be considered in setting cleanup goals). ACLs are derived through well-established procedures based on toxicity and exposure routes. Site usage and access to groundwater are important considerations.

Frequently, ACLs are less stringent than MCLs, allowing higher concentrations of contaminants to remain in place.

Another situation in which cleanup targets other than MCLs may be used is when groundwater is not suitable for drinking. EPA has a groundwater classification system that identifies groundwater as

- class I: special groundwater (irreplaceable source or potential source of drinking water),
- class II: groundwater currently or potentially a source of drinking water, or
- class III: groundwater not a source of drinking water due to insufficient yield, high salinity, or contamination that cannot be reasonably treated.

When groundwater fits class III, cleanup to goals other than MCLs is allowable.

Substantial variability exists among EPA regions and individual case managers in allowing the use of groundwater cleanup goals other than MCLs. Additionally, there are large differences in cleanup levels and types of groundwater classifications among individual states. At DOE sites, both the state within which the site is located and the EPA have influence over the required cleanup levels. Nonetheless, despite the occasional use of ACLs and the classification of some groundwater as nonpotable, MCLs have historically served as baseline standards for groundwater cleanup at RCRA and CERCLA sites.

EPA has established a separate set of groundwater cleanup targets under UMTRCA. Table 2-4 lists these requirements. For contaminants for which no goal has been established under UMTRCA, cleanup must achieve background levels. Alternate concentration limits are also allowable if DOE determines that the contaminants "will not pose a substantial present or potential hazard to human health and the environment as long as the alternate concentration limit is not exceeded," according to UMTRCA regulatory documents (40 C.F.R. 192).

Soil Baseline Cleanup Goals

For contaminated soils, there is no ARAR equivalent to the drinking water MCL. Until recently, soil cleanup goals were negotiated on a case-by-case basis, which increased the time to develop cleanup goals and costs and resulted in cleanup requirements that varied with location. Recognizing this limitation, EPA (1996) developed soil screening guidance for the establishment of cleanup levels. The soil screening guidance provides a tiered approach to estimate soil screening levels (SSLs) that may serve as preliminary remediation goals under certain

TABLE 2-4 Groundwater Cleanup Standards for UMTRCA Sites

Contaminant	Maximum Allowable Concentration (mg/liter, unless otherwise indicated)
Arsenic	0.05
Barium	1.0
Cadmium	0.01
Chromium	0.05
Lead	0.05
Mercury	0.002
Selenium	0.01
Silver	0.05
Nitrate (as nitrogen)	10
Molybdenum	0.1
Combined radium-226 and radium-228	5 pCi/liter
Combined uranium-234 and uranium-238	30 pCi/liter
Gross alpha-particle activity (excluding radon and uranium)	15 pCi/liter
Endrin	0.0002
Lindane	0.004
Methoxychlor	0.1
Toxaphene	0.005
2,4-Dichlorophenoxyacetic acid	0.1
2,4,4-Trichlorophenoxypropionic acid	0.01

SOURCE: 40 C.F.R. 192.

conditions. The soil screening framework considers potential exposures from ingestion of soil, inhalation of volatile compounds and fugitive dusts, and ingestion of groundwater contaminated by migration of chemicals through the soil to an underlying drinking water aquifer. SSLs are generally based on a 10^{-6} risk (meaning one excess cancer death per million people) for carcinogens or a hazard quotient of 1 (the exposure concentration divided by the "safe" dose) for noncarcinogens. SSLs apply at the point of potential exposure. For groundwater pathways, SSLs are back-calculated using drinking water standards at the site boundary.

Points of Compliance

Under CERCLA, groundwater cleanup goals must be achieved throughout the contaminated site, with the exception of underneath "waste management" areas. According to the National Contingency Plan (NCP), which is the primary EPA regulatory document for implementing CERCLA,

For groundwater, remediation levels should generally be attained

> throughout the contaminated plume, or at and beyond the edge of the waste management area when waste is left in place (EPA, 1990).

According to this policy, waste management areas are landfills that will be contained when remediation of the rest of the site is completed. EPA guidance documents indicate that DNAPLs that remain in place are not considered waste management areas (EPA, 1996). If DNAPLs cannot be cleaned up due to technical limitations, then ARARs should be waived, rather than changing the point of compliance for cleanup, according to EPA guidance documents (EPA, 1996). The NCP allows for changes in points of compliance under specific circumstances, including situations in which "there would be little likelihood of exposure due to the remoteness of the site . . . provided contamination in the aquifer is controlled from further migration" (EPA, 1990).

Points of compliance for soil cleanup under CERCLA are determined on a site-specific basis. In general, any soil containing concentrations of contaminants above the predetermined cleanup standards for the site must be treated. Containment of contaminated soil onsite is permissible, but CERCLA indicates that treatment should be attempted for "hot spots" containing high concentrations of contaminants. Where contaminated soil is contained, the point of compliance is the edge of the containment system.

According to EPA policy, points of compliance at RCRA sites are to be determined according to the same standards as points of compliance at CERCLA sites. In some instances, however, RCRA regulatory managers at specific sites have not followed the CERCLA policy, and the property boundary has been used as the point of compliance (K. Lovelace, Environmental Protection Agency, personal communication, 1998). This difference may be due to the fact that EPA generally empowers state agencies to implement RCRA, whereas EPA itself oversees CERCLA cleanups.

Like CERCLA and RCRA sites, the point of compliance at UMTRCA sites is the edge of any location in which remaining waste is to be contained in place. According to the policy for UMTRCA cleanups, the point of compliance is located "at the hydraulically downgradient limit of the disposal area plus the area taken up by any liner, dike, or other barrier designed to contain the residual radioactive material" (40 C.F.R. 192.02[c]).

Cleanup Goals at DOE Installations

Groundwater and soil cleanup goals at DOE installations must satisfy the requirements of applicable regulations, which in most cases means regulations under CERCLA and RCRA. Box 2-2 shows ground-

BOX 2-2
Example Baseline Cleanup Goals at DOE Sites

The following are sample cleanup goals established for select sites at a few of the major DOE installations. Information about cleanup goals at each of these sites was obtained from the ROD for the installation.

Hanford

1100 Area, Operable Unit 1

This 13-km^2 site is located in an area of the Hanford installation known as the Arid Lands Ecology (ALE) reserve. The reserve is currently used for ecological research but formerly contained a NIKE missile base and control center. Sources of contamination in operable unit 1 of the site include a battery acid pit, paint solvent pit, antifreeze and degreaser pit, antifreeze tank, and landfill. Groundwater and soil are contaminated.

The groundwater cleanup goal for the site is to reduce trichloroethylene (TCE) contamination to less than 0.5 mg/liter via natural attenuation. The ROD states that this concentration "is based on SDWA MCLs," but the MCL for TCE is 0.005 mg/liter, well below the goal specified in the ROD.

Soil cleanup goals for the site are based on state standards and include benzene, 0.5 mg/kg; toluene, 40 mg/kg; xylenes, 20 mg/kg; perchloroethylene (PCE), 0.5 mg/kg; TCE, 0.5 mg/kg; polycyclic aromatic hydrocarbons, 1 mg/kg; PCBs, 5.2 mg/kg; hexavalent chromium, 1,600 mg/kg; and lead, 250 mg/kg.

200 Area, 200-ZP-1 Operable Unit

The 200 Area of Hanford, a 40-km^2 (15-mi^2) tract, contains several operable units where cleanup is occurring. The contamination in the 200-ZP-1 operable unit originated from discharges from Hanford's Plutonium Finishing Plant into three liquid waste disposal sites. Soil is contaminated with a mixture of carbon tetrachloride (CCl_4) and plutonium. Almost all of the plutonium is bound to the soil, and very little has reached the groundwater, but the groundwater contains a plume of CCl_4.

Final soil and groundwater cleanup goals for this site have not been established. Meanwhile, an interim remedial measure is in place to stop the spread of the CCl_4 plume in the groundwater by using a pump-and-treat system.

Idaho National Engineering and Environmental Laboratory

Test Area North

This site is a 26-km^2 (10-m^2) area built in the 1950s to support the Aircraft Nuclear Propulsion Program sponsored by the Air Force and the Atomic Energy Commission. An injection well was used to dispose of industrial and sanitary wastes and wastewaters from 1953 until 1972. Groundwater at the site is contaminated.

Cleanup goals for contaminated groundwater are based on MCLs and are as follows: TCE, 0.005 mg/liter; PCE, 0.005 mg/liter; lead, 0.05 mg/liter; and strontium-90, 300 pCi/liter.

continues on next page

Pit 9

This site is a 4,000-m^2 (1-acre) pit used to dispose of drums, boxes, and other large items between 1967 and 1969. Drums of waste, including sludge contaminated with a mixture of transuranic elements and organic solvents, and boxes containing empty contaminated drums from Rocky Flats, account for 3,100 m^3 of the waste. Monitoring wells have indicated that the groundwater beneath Pit 9 is not contaminated. However, contaminants are present in soil and debris.

Soil and debris cleanup goals for treated waste containing less than or equal to 100 pCi/g and being returned to Pit 9 are based on "maximum allowable leachate concentrations" for RCRA delisting and health risk-based levels: CCl_4, 18 mg/kg; PCE, 45 mg/kg; 1,1,1-trichloroethane (1,1,1-TCA), 2,910 mg/kg; TCE, 15 mg/kg; potassium cyanide, 119 mg/kg; and sodium cyanide, 122 mg/kg.

Soil and debris cleanup goals for treated waste residuals containing less than 10 pCi/g of radioactivity and being temporarily stored onsite are based on RCRA land disposal restrictions: CCl_4, 5.6 mg/kg; PCE, 5.6 mg/kg; 1,1,1-TCA, 5.6 mg/kg; TCE, 5.6 mg/kg; potassium cyanide, 122 mg/kg; lead, 5 mg/liter; and mercury, 260 mg/kg.

Oak Ridge Reservation, Operable Unit 16

This site includes two former disposal ponds used for waste from the former K-25 uranium enrichment facilities. DOE removed the remaining sludge from both ponds in 1987 and 1988 to comply with RCRA requirements for site closure and subsequently discovered contamination in the underlying soil and groundwater.

Groundwater cleanup goals and a groundwater remediation plan for the site have yet to be established, but soil cleanup goals have been determined. According to the ROD, the soil cleanup goals are based on "a health-risk level of 10^{-6}, EPA-recommended equations for calculating preliminary remediation goals for radionuclides in soil, and RCRA clean closure requirements." The soil cleanup goals include the following: americium-241, 0.002 pCi/g; cadmium, 1 mg/kg; cesium-137, 0.004 pCi/g; chromium, 0.000002 mg/m^3; cobalt-60, 0.002 pCi/g; europium-154, 0.004 pCi/g; manganese, 156 mg/kg; mercury, 0.1 mg/kg; neptunium-237, 0.002 pCi/g; nickel, 130 mg/kg; potassium-40, 0.033 pCi/g; technetium-99, 1.8 pCi/g; thorium-230, 0.003 pCi/g; uranium-234, 0.003 pCi/g; uranium-235, 0.007 pCi/g; uranium-238, 0.001 pCi/g; and zinc, 52 mg/kg.

Savannah River Site

All groundwater contamination sites mentioned in the ROD for the Savannah River Site either (1) require no further action under CERCLA because contamination is being managed under RCRA or (2) have no specified cleanup goals because at this time only interim actions have been determined.

water and soil cleanup standards for a sampling of sites from DOE installations, based on a review of RODs for these installations. The box includes information from sites where cleanup goals have not yet been specified. An informal review of RODs indicates that goals have yet to be determined for most of the DOE sites.

CHANGING REGULATORY ENVIRONMENT

Although baseline cleanup standards generally must be achieved across most of the area of a contaminated site, changes in these standards and in the overall process of site cleanup under RCRA and CERCLA are becoming increasingly common. The driver for some of these changes is recognition of the limits of available technologies for site cleanup: achieving existing baseline standards is not possible for certain types of contamination scenarios with existing technologies (NRC, 1994). The driver for other changes in regulatory practice is recognition of the extremely high costs of cleanup using conventional methods. SCFA managers need to be aware of this changing regulatory environment because it opens up new possibilities for site cleanup at DOE installations. Technologies that are unable to achieve baseline standards or that regulators might not have considered in the past may be acceptable for use in the new regulatory environment.

Five broad emerging trends, discussed below, show how the nature of contaminated site regulation is changing: (1) increasing interest in the use of "technical impracticability waivers" where groundwater restoration is not technologically feasible; (2) increasing use of monitored natural attenuation (intrinsic remediatioÙ! in place of engineered cleanup systems; (3) increasing number of changes to groundwater and soil remedies specified in CERCLA RODs; (4) emergence of brownfields programs allowing site cleanup to industrial reuse standards rather than residential standards; and (5) emergence of risk-based programs at the state level for assessing site cleanup requirements.

Technical Impracticability Waivers

Under both CERCLA and RCRA, required cleanup standards for contaminated groundwater and soil can be waived in cases where achieving these standards is not possible with existing technologies. For example, CERCLA states that cleanup standards can be waived if cleanup is "technically impracticable from an engineering perspective" (EPA, 1990). RCRA contains similar language. Both statutes state that engineering feasibility and reliability, rather than cost, should be the key considerations in determining the practicality of cleanup.

DOE's strategy for cleaning up its installations (known as the "Paths to Closure" Plan) emphasizes that because of the limitations of existing remediation technologies, groundwater contamination will remain at many sites after other cleanup goals are achieved (DOE, 1998). If this is the case, DOE will have to apply for waivers to cleanup standards at many sites, and DOE's cleanup managers will have to be familiar with EPA's policies concerning cases where groundwater restoration is technically infeasible. Further, SCFA managers will have to be aware of what remedial alternatives are acceptable to EPA in cases where achievement of cleanup standards is infeasible because these will affect which technologies should be selected for development.

Prior to the early 1990s, regulatory policies for implementing the "technical impracticability" provisions of CERCLA and RCRA were ambiguous. In 1993, due to increasing recognition of the limitations of groundwater cleanup technologies, EPA issued a guidance document clarifying its policies on granting waivers to cleanup standards based on technical considerations and specifying how site owners should go about applying for such waivers (EPA, 1993). EPA intended that this guidance document, titled *Guidance for Evaluating the Technical Impracticability of Ground-Water Restoration*, would provide a consistent standard for use by EPA staff overseeing CERCLA and RCRA sites in deciding whether or not cleanup of groundwater is technically feasible. The cover memo (signed by the acting administrator of the CERCLA and RCRA programs) accompanying the guidance document states that "experience over the past decade has shown that achieving the required final cleanup standards may not be practicable at some sites due to the limitations of remediation technology" (Guimond, 1993).

Although the guidance document is applicable to all types of contaminants, it emphasizes DNAPLs. The guidance document states, "As proven technologies for the removal of certain types of DNAPL contamination do not exist yet, DNAPL sites are more likely to require TI [technical impracticability] evaluations than sites with other types of contamination." It indicates that up to 60 percent of CERCLA sites may contain DNAPLs, according to EPA surveys.

The key elements of the technical impracticability guidance document are its discussion of (1) the timing of decisions concerning the technical impracticability of cleanup, (2) alternative remedial strategies where cleanup is not possible, (3) long-term monitoring requirements for sites where cleanup goals are waived due to technical impracticability, and (4) types of data that must be provided to EPA to evaluate the technical feasibility of achieving cleanup standards.

Regarding timing, the guidance document specifies that in many cases, EPA staff should hold off on granting technical impracticability waivers until a full-scale cleanup system has been implemented and has failed to achieve cleanup standards. The document states, "EPA believes that, in many cases, TI decisions should be made only after

TABLE 2-5 Technical Impracticability Waivers in CERCLA RODs for Sites with Contaminated Groundwater, 1989-1997

Year	Number of Waivers	Number of RODs for Groundwater	Percentage of RODs with Waivers
1989	4	83	4.8
1990	2	109	1.8
1991	7	135	5.2
1992	4	93	4.3
1993	2	82	2.4
1994	2	96	2.1
1995	4	89	4.5
1996	1	93	1.1
1997	3	NA	NA
TOTAL	29		

NOTE: NA indicates that data are not available.

SOURCE: K. Lovelace, EPA, unpublished data, 1998.

interim or full-scale aquifer remediation systems are implemented because often it is difficult to predict the effectiveness of remedies based on limited site characterization data alone." It specifies that technical impracticability waivers can be granted prior to trying a full-scale remedy only "in cases where there is a high degree of certainty that cleanup levels cannot be achieved." Data from the CERCLA program indicate that, in general, EPA is following the policy of granting impracticability waivers in most cases only after full-scale remedies have been installed. Table 2-5 shows the number of technical impracticability waivers specified in RODs between 1989 and 1997; as shown, the number of sites with RODs specifying technical impracticability is still very small and has remained at a relatively constant level. Presumably, the number of such waivers is small because waivers are not generally granted without attempting cleanup. This requirement may pose difficulties for DOE in obtaining technical impracticability waivers, because full-scale groundwater cleanup systems are not yet in place for most sites.

The guidance also specifies that technical impracticability waivers must include alternative remedial strategies to protect public health and the environment when cleanup standards cannot be achieved. The alternative remedial strategy must document how exposure to the contamination will be prevented (for example, through restrictions on well construction), how the source of contamination will be controlled, and how plumes of dissolved contaminants emanating from source areas will be managed. It suggests well construction and deed restrictions as possible exposure control methods. It also indicates that contaminant sources should be contained either hydraulically or physically to the maximum extent possible. Where sources are effectively contained, it requires that plumes of dissolved contaminants be restored to applicable standards. Where source containment is not possible, either plumes should be controlled hydraulically, less stringent cleanup levels can be established, or contaminants can be allowed to attenuate naturally, as long as exposure controls are in place. EPA's general preference for source containment may require that SCFA strengthen its efforts to help develop effective long-term containment systems.

The guidance document describes in detail the types of supporting information that site owners must provide when applying for technical impracticability waivers. Essentially, applications for such waivers must include the following five parts:

1. identification of required cleanup standards for which technical impracticability waivers are being sought;
2. description of the spatial area for which the waiver is being sought;
3. conceptual model showing site geology, hydrology, groundwater contamination sources, and contaminant transport and fate;
4. evaluation of the site's "restoration potential," to include
   - proof that contamination sources have been identified and will be removed or contained to the extent possible,
   - analysis of performance of any existing remediation systems,
   - predicted time to attain required cleanup levels with available technologies, and
   - evidence that no existing technology can attain required cleanup levels within a reasonable time period; and
5. estimates of the cost of existing or proposed remedy options.

Thus, for sites at which DOE believes that groundwater cleanup is not technically feasible, DOE managers will have to supply EPA with the above types of information.

One of the most important provisions of the technical impracticability guidance document is EPA's right to require additional work in future years at sites with technical impracticability waivers. Sites with such waivers remain "open" to future requirements by EPA. The document specifies that EPA will reassess CERCLA sites with such waivers every five years and will reassess RCRA sites periodically, as well. If new technologies emerge that might restore the groundwater, EPA can require that they be implemented. EPA will

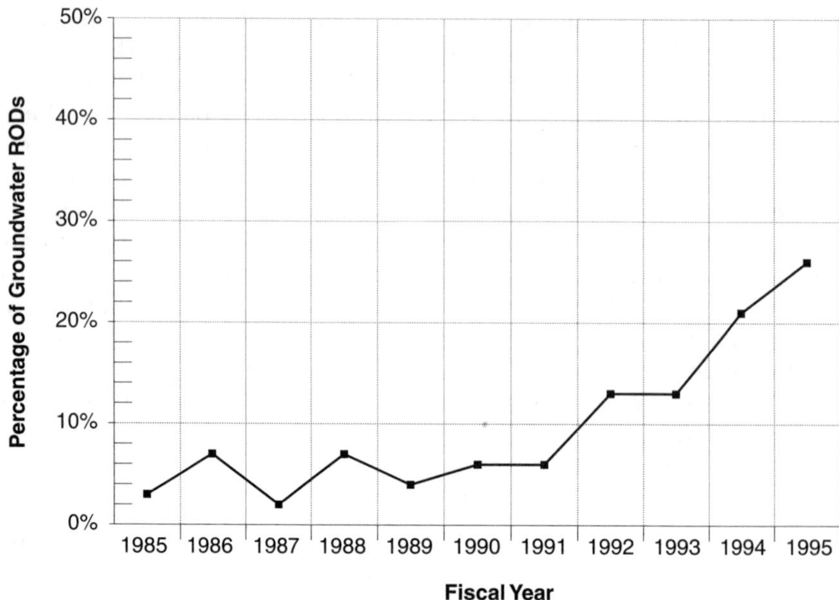

FIGURE 2-3 Use of natural attenuation in the cleanup of contaminated groundwater at CERCLA sites, 1985-1995. SOURCE: K. Lovelace, EPA, unpublished data, 1998.

also require continuous, long-term monitoring of these sites. Technical impracticability waivers therefore will not eliminate DOE's long-term liability for remaining contamination.

Monitored Natural Attenuation

Over the past two decades, a body of knowledge has accumulated indicating that some types of contaminants, especially petroleum hydrocarbons, can degrade naturally in the subsurface at relatively rapid rates (NRC, 1993). This knowledge is increasingly reflected in the practice of cleaning up contaminated sites. To a greater degree, regulators are approving the use of "natural attenuation," rather than engineered cleanup remedies, to solve groundwater contamination problems or reduce the size of the area treated with engineered remedies. Figure 2-3 shows the increase in the use of natural attenuation for contaminated groundwater at CERCLA sites between 1985 and 1995. DOE managers need to be aware of current regulatory policies concerning the use of natural attenuation in order to determine whether this strategy may apply at some of their sites. SCFA has to be aware of these policies because the use of natural attenuation at DOE sites will require additional research for some types of contaminants and the development of better tools for monitoring the progress of natural attenuation.

The key EPA policy document pertaining to natural attenuation is a directive entitled *Use of Monitored Natural Attenuation at Superfund, RCRA Corrective Action, and Underground Storage Tank Sites*, finalized in 1999 (EPA, 1999). The directive codifies that natural attenuation can be an acceptable remedy for contaminated groundwater and soil at CERCLA and RCRA sites. Publication of this directive marked a change from the policies of the 1980s and early 1990s. In the earlier years, strong preference was given to engineered remedies (Brady et al., 1997), and natural attenuation was rarely used, as shown in Figure 2-3.

The directive defines monitored natural attenuation as "the reliance on natural attenuation processes . . . to achieve site-specific remediation objectives within a time frame that is reasonable to that offered by other more active methods." According to the directive, the natural processes that can contribute to contaminant attenuation include biodegradation, dispersion, dilution, sorption to solid media, volatilization, and chemical or biological stabilization, transformation, or destruction.

The directive indicates that in order to apply for the use of monitored natural attenuation, site owners, with certain exceptions, must submit the following types of data:

1. historical groundwater and soil data showing a continuous trend of decreasing contaminant concentration or decreasing contaminant mass over time;
2. hydrogeologic and geochemical data demonstrating indirectly the types of natural attenuation processes at work at the site and the rate at which these processes will reduce contaminant concentrations; and
3. data from microcosm studies conducted with contaminated media from the site directly proving that specific processes are active at the site.

The directive indicates that the first type of data will always be required. EPA will require the second type of data in all cases except when "EPA or the overseeing regulatory authority determines that historical data (number 1 above) are of sufficient quality and duration to support a decision to use MNA [monitored natural attenuation]." The third type of data will be required only where the first two types of data are "inadequate or inconclusive," according to the directive. The directive does not contain specific guidance on what types of data are appropriate for these three categories or on how to gather such data.

The directive provides several caveats about the use of monitored natural attenuation that site managers must consider when assessing whether to apply for regulatory approval to use this remediation strategy. Key among the caveats are the following:

- Active measures to control contaminant sources usually will be required even at sites where monitored natural attenuation is approved.
- Site characterization data for natural attenuation will have to be more detailed than for other remedies.
- Performance monitoring data for natural attenuation must be more detailed than for other remedies.
- The time frame for natural attenuation should not be excessively long compared to the time frame for engineered remedies.
- Contingency remedies will have to be specified in the event that natural attenuation fails.

Implementing natural attenuation remedies at DOE sites may thus require additional research to determine how to provide the three categories of data mentioned above, how best to control contaminant sources, how to monitor the site, and how best to predict the likely time frame for natural attenuation. Research on these topics is currently under way at some of DOE's national laboratories (see, for example, Brady et al., 1998a,b).

EPA and a variety of other organizations have also developed technical documents indicating how to assess sites for natural attenuation potential. The most detailed of these, providing guidelines on how to gather appropriate site data, were developed by the Air Force Center for Environmental Excellence and cover petroleum hydrocarbon and chlorinated solvent contamination (Wiedemeier et al., 1995, 1997). The EPA recently released a technical protocol similar to the Air Force protocol for evaluating natural attenuation potential at sites contaminated with chlorinated solvents (EPA, 1998). Other organizations that have developed natural attenuation protocols include the Navy (Wiedemeier and Chapelle, 1998); the American Society for Testing and Materials (1997); a few states (see, for example, Minnesota Pollution Control Agency, 1997); the Remediation Technologies Development Forum (1997); the American Petroleum Institute (1997); and Chevron Corporation (Buscheck and O'Reilly, 1995, 1997). DOE researchers at Sandia National Laboratories also recently developed a document entitled *Site Screening and Technical Guidance for Monitored Natural Attenuation at DOE Sites* (Brady et al., 1998a). Some states also have their own policies and guidance documents for natural attenuation. EPA's monitored natural attenuation guidance, however, cautions that its own policy, rather than these other documents, will provide the basis for approving the use of monitored natural attenuation. It states that non-EPA natural attenuation guidance manuals are not "officially endorsed by the EPA, and all parties should clearly understand that such guidances do not in any way replace current EPA . . . guidances or policies addressing the remedy selection process."

Changes to Records of Decision

Until the mid-1990s, changes to CERCLA RODs in order to allow the use of a more effective technology were extremely rare. The inability to change to a different remediation technology once the ROD had been signed created a barrier to the use of innovative remediation technologies (NRC, 1997), both at DOE sites and elsewhere. Years can pass between signing of the ROD and construction of the cleanup remedy (Guerrero, 1998). During this time, new technologies may emerge that could improve the prospects for site cleanup or reduce costs.

Since 1995, EPA has changed its policies concerning ROD revisions and increasingly is allowing modifications to remedies specified in RODs to reflect new information about cleanup technologies or new understanding about the site. DOE and SCFA managers must be familiar with these policy changes because the new policies open an avenue for increasing use of innovative, cost-saving remedies at DOE sites. In 1996 and 1997, EPA approved remedy changes for groundwater and/or soil in existing RODs at 130 CERCLA sites. As shown in Table 2-6, the greatest number of changes (35) were approved to allow modifications to the design of the original remedy, often to reflect new performance data. Conventional remedies were changed to innovative remedies at another 11 sites. At nine sites, conventional pump-and-treat systems were eliminated and changed to monitored natural attenuation. Changes in required cleanup levels (usually to less stringent levels) were also allowed at a number of sites.

In addition to allowing more flexibility in changes to remedies once RODs are signed, EPA has instituted a formal program to review all planned high-cost remedies, either just before the ROD is signed or (in a few cases) after the ROD is signed (Laws, 1995; Luftig, 1996). These reviews are carried out by the National Remedy Review Board, formed in January 1996. The review board formally assesses planned remedies for all non-DOE sites for which the action costs more than $30 million or for which the remedy costs more than $10

TABLE 2-6 Changes in CERCLA RODs for Groundwater and Soil Contamination, 1996-1997

Reason for Change	Number of Sites
Minor modification of original remedy design[a]	35
Treatment of full site changed to treatment of hot spots, or boundaries of remedy area decreased	5
Boundaries of remedy area increased	3
Change in required cleanup levels, intended land use, number of contaminants covered, or regulatory authority	20
Conventional remedy changed to innovative remedy	11
Innovative remedy changed to poor or uncertain performance	8
Pump-and-treat system changed to monitored natural attenuation	9
Pump-and-treat system changed to containment and/or monitoring	5
Pump-and-treat system downsized	5
In situ soil treatment or containment changed to ex situ treatment	6
Soil remedy changed to capping and/or containment	8
On-site soil treatment changed to off-site disposal or treatment	12
Off-site treatment or disposal changed to on-site treatment or disposal	3
Cleanup goals achieved; treatment discontinued	3
Change in duration of monitoring	1
Other	2

NOTE: Categories were based on interpretation of data provided by M. Charsky, EPA, 1998. Total number of sites with ROD changes is 130. Changes occurred for more than one reason at some sites.

[a]Modifications include changes in treatment system for extracted water, location for soil disposal, design of incinerator, design of landfill, and others.

million and is 50 percent more costly than the least-cost alternative that can meet cleanup criteria for the site. For DOE sites, the thresholds for review are $75 million and $25 million, respectively. The review board consists of managers and senior scientific staff from EPA headquarters in Washington and the regional offices.

Following its review of a high-cost site, the review board issues recommendations to EPA decision makers in the region in which the site is located. The decision makers are not required to adopt the board's recommendations but must, at a minimum, prepare a written response indicating the logic of the choice to address or not address the board's concerns. The board reviewed 23 cleanup decisions, including two at DOE's Fernald facility and one at Oak Ridge National Laboratory, between its formation and January 1998 (NRRB, 1998).

The existence of this board may trigger the search for more cost-effective solutions for high-cost sites; it may prompt site managers to seek new technologies that will reduce cleanup costs to levels below the board's threshold cost criteria so that they can avoid having to undergo the formal review process.

Brownfields Programs

Although historically the goal of contaminated site cleanup programs has been to return sites to conditions that would allow unrestricted future use, increasing numbers of sites are now being cleaned up to levels safe for industrial and commercial use but not for residential use. Many of these sites are being restored under the auspices of "brownfields" programs.

Brownfields are former commercial and industrial facilities that have been idled due in part to contamination problems and that could be returned to productive use provided some of the contamination can be removed. In general, sites cleaned up under brownfields programs can be converted to new uses much more rapidly than those restored under CERCLA and RCRA because the regulatory process is much less cumbersome and because cleanup standards are scaled back to provide protection for commercial and industrial users but not for potential residential users of the property. Some DOE installations, especially those that are near metropolitan areas and that DOE would like to turn over to the private sector, could be restored under brownfields programs. DOE and SCFA managers need to be familiar with the scope of these programs because their cleanup goals typically differ from those of CERCLA and RCRA, and thus the suite of possible remediation approaches differs as well.

More than 200 contaminated sites across the country are now receiving funding from EPA for brownfields cleanups under the President's Brownfields Initiative, launched in November 1993 (EPA Region 8, 1998). Under this initiative, site owners or affected communities can apply for grants of up to $200,000 to serve as seed money for brownfields cleanups. Past projects have used this funding for gathering more detailed site characterization data to clarify the nature and extent of contamination, preparing redevelopment plans, setting cleanup priorities, and establishing working relationships with concerned citizens. EPA's brownfields funds can be used for federal facilities, such as DOE installations, as well as for private-sector sites.

Many state and local governments have also created special programs to encourage redevelopment of brownfields (GAO, 1995). Key to these programs is reducing the fear of future liability for contamination. In the past, the resale and reuse of idle contaminated property have been hindered by fear that the new owner or a lender who provides funding for the owner will be held liable for contamination discovered onsite, even if the owner and lender were not responsible for the contamination. Many state and local governments have established mechanisms to protect purchasers and lenders against liability; these mechanisms include special legislation and written covenants not to sue. Also available through some of these programs are loans for redevelopment.

Key to brownfields programs is a formal change in the allowable land use for the site once cleanup has been completed. Case studies have shown that land use provisions have enormous effects on the residual risk at a site once a site is cleaned up. For example, Katsumata and Kastenberg (1997) demonstrated that at one CERCLA site, scenarios that assumed future residential use of the site produced risks from one to three orders of magnitude greater than scenarios assuming continued industrial use of the property. This difference was due to different assumptions about where and how humans would be exposed to the contamination. Further, they demonstrated that the planned cleanup remedy for the site (involving excavation of the soil and pumping and treating of groundwater) would not reduce risk sufficiently to protect future residential users even though the remedy was intended to do so; remaining risks would exceed EPA's general allowable threshold of one excess cancer case per 10,000 residents. The planned remedy would, however, be sufficient to reduce risks to below the 1-in-10,000 threshold for workers at a future industrial facility located on the site. The suite of acceptable remediation technology alternatives is thus likely to be broader at brownfield sites than at sites restored to residential use standards. SCFA may want to consider alternative possible cleanup end points and the availability of technologies that can achieve these end points in planning its remediation technology development program.

Involvement in brownfields programs is a potentially important component of DOE's remediation strategy for some of its contaminated installations. DOE has had limited involvement in brownfields programs so far. DOE is a member of the Interagency Working Group on Brownfields (established in 1996 as a forum for information exchange among federal agencies) and in 1997 provided $315,000 to begin working with communities at DOE installations in potential brownfields areas (EPA, 1997a).

Risk-Based Corrective Action Programs

An increasing number of state environmental agencies are adopting a process known as "risk-based corrective action" (RBCA) to evaluate and select cleanup remedies for sites that the states oversee. The RBCA process was developed by the American Society for Testing and Materials (ASTM) and published in the form of two industrial standard guides (ASTM, 1995, 1998). The first standard guide, published in 1995, applies to sites contaminated with petroleum hydrocarbons. The second, published in 1998, applies to sites with other chemical contaminants. Most states allow use of the RBCA process or a variant. As of October 1998, 14 states had formally adopted ASTM's RBCA standard as part of their regulatory process for petroleum-contaminated sites, and 27 additional states were developing RBCA programs (S. McNeely, EPA Office of Underground Storage Tanks, personal communication, 1998). In addition, the Air Force has developed a methodology, the Enhanced Site Specific Risk Assessment process, similar to RBCA and is conducting an in-depth feasibility analysis to examine how the methodology will fit into the cleanup process. The Navy is considering using RBCA or a similar process for cleaning up its sites, as well. DOE cleanup managers and SCFA should be familiar with the RBCA process because its use may eventually expand beyond the cleanup of petroleum-contaminated sites, and it therefore may influence the selection of cleanup remedies.

RBCA integrates site assessment, remedy selection, and site monitoring through a tiered approach involving increasingly sophisticated levels of data collection and analysis (see Figure 2-4). The initial site assessment identifies source areas of chemicals of concern, potential human and environmental receptors, and potentially significant transport pathways. Sites are then classified and initial response actions identified based on the urgency of need (immediate, zero to two years, more than two years, no action). Based on the information obtained during the initial site assessment, project managers perform a "tier 1" evaluation (according to steps specifically outlined in the RBCA standard) to determine whether the site qualifies for quick regulatory closure or warrants a more site-specific evaluation. In determining risk, the tier 1 evaluation uses standard exposure scenarios with current reasonable maximum exposure assumptions and toxicological parameters. When the tier 1 evaluation indicates a possible risk to human health, project managers can decide to clean up the site or proceed to a more detailed site risk evaluation, known as tier 2. At the end of tier 2, project managers again have the option of closing

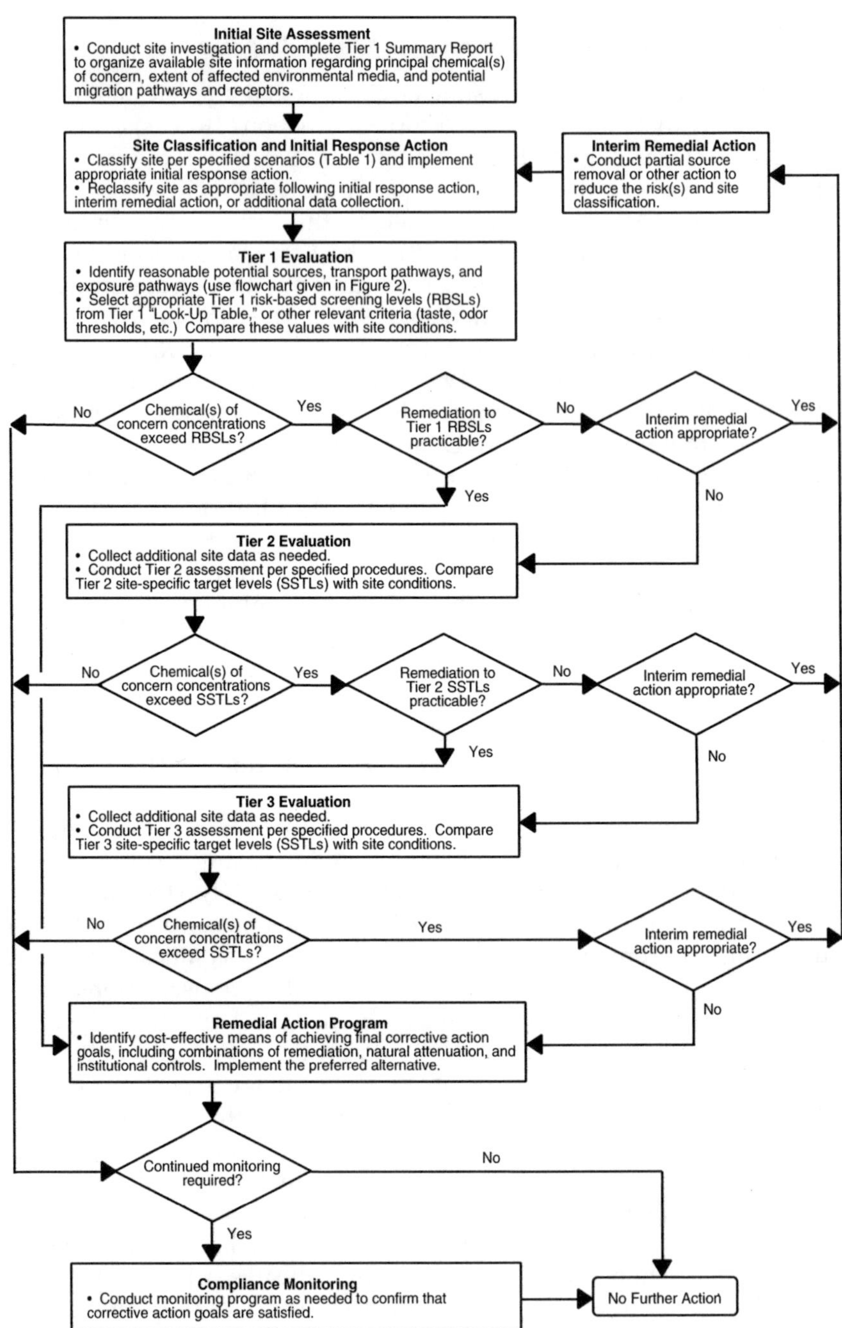

FIGURE 2-4 Steps in the ASTM RBCA process. SOURCE: ASTM, 1995.

the site (if the more detailed evaluation shows that there is no risk), cleaning up the site to protect against risks as computed in tier 2, or proceeding to a final level of highly detailed evaluation, known as tier 3. Tier 3 provides the flexibility for more complex calculations to establish cleanup levels and may include additional site assessment, probabilistic evaluations, and sophisticated chemical fate and transport models.

In a recent review of application of RBCA and other risk-based approaches for cleanup of Navy installations, the National Research Council (NRC, 1998) concluded that the RBCA approach has several advantages, including the following:

- Site assessment activities can be focused on collecting only information that is necessary to make risk-based corrective action decisions.
- Remedial decisions may be accelerated and costs therefore reduced.
- Resources can be focused on sites that pose the greatest risk to human health.

However, the NRC also concluded that the existing RBCA standards would not be suitable for application at Navy sites because the standards lack 6 of 11 criteria that the NRC determined are essential for the successful implementation of risk-based approaches at Navy facilities. In particular, the RBCA standards are lacking in the following:

1. They do not provide for integrated assessment of multiple sites affecting the same human or ecological receptors at the same installation.
2. They lack sufficient mechanisms for considering critical uncertainties in site assessment, such as those associated with models of contaminant fate and transport and with estimating health and ecological risks.
3. They do not adequately account for long-term risks that may remain in place even after cleanup has been completed to the extent practicable.
4. They do not adequately address the need for public involvement in remedy selection.
5. They did not undergo external, independent scientific peer review and public review.
6. It is not clear whether the standards can satisfy all of the regulatory requirements under Superfund and RCRA.

The NRC recommended that the Navy develop a risk-based methodology for its Environmental Restoration Program that satisfies 11

BOX 2-3
NRC's Criteria for a Risk-Based Remediation Process

In a 1998 review of the potential for application of risk-based methodologies in the cleanup of contaminated Navy bases, the NRC concluded that in order to succeed, any risk-based approach must meet the following criteria:

1. It facilitates prioritization of contaminated sites at individual installations.
2. It provides a mechanism for increasing the complexity of the remedial investigation when appropriate.
3. It provides guidance on data collection needed to support the development of site-specific cleanup goals.
4. It provides for integrated assessment of sites affecting the same human or ecological receptors.
5. It encourages early action at sites (1) where the risk to human health and the environment is imminent and (2) for which the risks are demonstrably low and remediation is likely to be more rapid and inexpensive.
6. It considers relevant uncertainties.
7. It provides a mechanism for integrating the selection of the remedial option with the establishment of remedial goals. It also provides quantitative tools for developing risk management strategies.
8. It has options to revisit sites over the long term.
9. It is implemented in a public setting with all stakeholders involved.
10. It undergoes both external, independent scientific peer review and public review.
11. It complies with relevant state and federal statutory programs for environmental cleanup.

SOURCE: NRC, 1998.

basic criteria (see Box 2-3). Such criteria might also be applied to the use of risk-based cleanup processes at DOE installations.

CONCLUSIONS

An understanding of DOE's changing legal obligations for the cleanup of contaminated groundwater and soil is critical to the effective administration of SCFA's program for developing new groundwater and soil cleanup technologies. SCFA must tailor its technology development program to ensure that DOE has the tools necessary to meet applicable legal requirements. Although groundwater and soil restoration goals have not yet been specified for many DOE sites, when these goals are established they must satisfy the requirements of applicable regulations, generally RCRA, CERCLA, UMTRCA, or a combination of these.

Policies concerning cleanup requirements for groundwater and soil are evolving rapidly. SCFA managers need to keep track of these changes because new policies may affect the selection of cleanup goals for DOE sites and, correspondingly, the suite of possible remediation technologies for achieving those goals. Key trends for SCFA to monitor include the following:

- **Increasing use of technical impracticability waivers.** EPA has clarified its policies for determining when cleanup to baseline standards is infeasible. SCFA has to plan for the development of remedial alternatives (including containment systems) that can be used at sites where technical impracticability waivers are granted. Further, SCFA has to continue to pursue development of technologies that can clean up these difficult sites because issuance of a waiver does not remove DOE's long-term liability for a site.
- **Increasing use of monitored natural attenuation.** Use of monitored natural attenuation in place of or in conjunction with active cleanup remedies is increasing at contaminated sites nationwide, but implementing natural attenuation at DOE sites may require additional research to develop methods for predicting the fate of contaminants under conditions of natural attenuation. SCFA must understand current policy requirements for implementation of monitored natural attenuation and determine what additional research will be necessary for DOE to meet these requirements at appropriate sites.
- **An increase in the number of changes to groundwater and soil remedies in CERCLA RODs.** At an increasing number of CERCLA sites, remediation technologies specified in RODs are being changed to reflect new technological developments or new understanding about the site. SCFA could play a useful role in determining where innovative technologies might provide more effective solutions than technologies specified in current RODs.
- **Emergence of brownfields programs.** Increasing numbers of sites in former industrial areas are being cleaned up to industrial reuse standards, rather than residential use standards, under brownfields programs. DOE is eligible for participation in these programs, and SCFA should keep track of the types of technologies that might be appropriate for remediation under brownfield scenarios.
- **Emergence of risk-based corrective action programs.** Increasing numbers of organizations are developing risk-based procedures designed to set cleanup priorities among contaminated sites. If DOE managers decide such a process is appropriate for their sites, SCFA could play a role in developing the protocols for DOE. SCFA should be familiar with existing risk-based corrective action procedures developed by other organizations.

REFERENCES

American Petroleum Institute. 1997. Methods for Measuring Indicators of Intrinsic Bioremediation: Guidance Manual. Washington, D.C.: American Petroleum Institute.

ASTM (American Society for Testing and Materials). 1995. Standard Guide for Risk-Based Corrective Action at Petroleum Release Sites (E1739-95). Annual Book of ASTM Standards. West Conshocken, Pa.: ASTM.

ASTM. 1997. Standard Guide for Remediation of Groundwater by Natural Attenuation at Petroleum Release Sites: Draft. February 4. Philadelphia: ASTM.

ASTM. 1998. Standard Provisional Guide for Risk-Based Corrective Actions (PS104-98). Annual Book of ASTM Standards. West Conshohocken, Pa.: ASTM.

Brady, P. V., M. V. Brady, and D. J. Borns. 1997. Natural Attenuation: CERCLA, RBCA's and the Future of Environmental Remediation. Boca Raton, Fla.: Lewis Publishers.

Brady, P. V., B. P. Spalding, K. M. Krupka, R. D. Waters, Pengchu Zhang, D. J. Borns, and W. D. Brady. 1998a. Site Screening and Technical Guidance for Monitored Natural Attenuation at DOE Sites. Albuquerque, N.M.: Sandia National Laboratories.

Brady, P. V., B. P. Spalding, K. M. Krupka, R. D. Waters, Penchus Zhang, and D. J. Borns. 1998b. Site Screening for Monitored Natural Attenuation with MNA Toolbox. Albuquerque, N.M.: Sandia National Laboratories.

Buscheck, T., and K. O'Reilly. 1995. Protocol for Monitoring Intrinsic Bioremediation in Groundwater. Richmond, Calif.: Chevron Research and Technology Company.

Buscheck, T., and K. O'Reilly. 1997. Protocol for Monitoring Natural Attenuation of Chlorinated Solvents in Groundwater. Richmond, Calif.: Chevron Research and Technology Company.

DOE (Department of Energy). 1998. Accelerating Cleanup: Paths to Closure. DOE/CM-0362. Washington, D.C.: DOE, Office of Environmental Management.

EPA (Environmental Protection Agency). 1990. National Oil and Hazardous Substances Pollution Contingency Plan; Final Rule. Federal Register 55(46).

EPA. 1993. Guidance for Evaluating the Technical Impracticability of Ground-Water Restoration: Interim Final. Directive 9234.2-25. Washington, D.C.: EPA, Office of Solid Waste and Emergency Response.

EPA. 1995. Technical Resource Document: Extraction and Beneficiation of Ores and Minerals, Vol. 5—Uranium. EPA 530-R-94-032. Washington, D.C.: EPA, Office of Solid Waste, Special Waste Branch.

EPA. 1996. Presumptive Response Strategy and Ex-situ Treatment Technologies for Contaminated Groundwater at CERCLA Sites. EPA 540-R-96-023. Washington, D.C.: EPA, Office of Solid Waste and Emergency Response.

EPA. 1997a. Brownfields National Partnership Action Agenda. EPA 500-F-97-090. Washington, D.C.: EPA, Office of Outreach and Special Projects. http:/www.epa.gov/swerosps/bf/html-doc/97aa_fs.htm.

EPA. 1998. Technical Protocol for Evaluating Natural Attenuation of Chlorinated Solvents in Groundwater. EPA/600/R-98/128. Washington, D.C.: EPA, Office of Research and Development.

EPA. 1999. Use of Monitored Natural Attenuation at Superfund, RCRA Corrective Action, and Underground Storage Tank Sites. Directive Number 9200.4-17p. Washington, D.C.: EPA, Office of Solid Waste and Emergency Response.

EPA Region 8. 1998. Brownfields Pilot Grant Announced for Missoula, MT. Denver, Colo.: EPA Region 8, Office of Communication and Public Involvement. http://www.epa.gov/unix0008/html/news/news98/brnfmis.html.

GAO (U.S. General Accounting Office). 1994. Nuclear Cleanup: Difficulties in Coordinating Activities Under Two Environmental Laws. GAO/RCED-95-66. Washington, D.C.: GAO.

GAO. 1995. Community Development: Reuse of Urban Industrial Sites. GAO/RCED-95-172. Washington, D.C.: GAO.

Guerrero, P. F. 1998. Superfund: Times to complete site listing and cleanup. Testimony before the Subcommittee on Finance and Hazardous Materials, Committee on Commerce, House of Representatives, February 4. GAO/T-RCED-98-74. Washington, D.C.: U.S. General Accounting Office.

Guimond, R. J. 1993. Memorandum. Transmittal of OSWER directive 9234.2-25: "Guidance for Evaluating the Technical Impracticability of Ground-Water Restoration." Office of Solid Waste and Emergency Response, Washington, D.C.: Environmental Protection Agency.

Katsumata, P. T., and W. E. Kastenberg. 1997. On the impact of future land use assumptions on risk analysis for Superfund sites. Journal of the Air and Waste Management Association 47(August):881-889.

Laws, E. P. 1995. Memorandum. November 28: Formation of National Superfund Remedy Review Board. Washington, D.C.: Environmental Protection Agency, Office of Solid Waste and Emergency Response.

Luftig, S. D. 1996. Memorandum. September 26: National Remedy Review Board. Washington, D.C.: Environmental Protection Agency, Office of Solid Waste and Emergency Response.

Minnesota Pollution Control Agency. 1997. Draft Guidelines: Natural Attenuation of Chlorinated Solvents in Groundwater. Working Draft. December 12. St. Paul, Minn.: Minnesota Pollution Control Agency.

NRC (National Research Council). 1993. Pesticides in the Diets of Infants and Children. Washington, D.C.: National Academy Press.

NRC. 1994. Alternatives for Groundwater Cleanup. Washington, D.C.: National Academy Press.

NRC. 1997. Innovations in Groundwater and Soil Cleanup: From Concept to Commercialization. Washington, D.C.: National Academy Press.

NRC. 1998. Assessing the Use of Risk-Based Methodologies for Environmental Remediation at Naval Facilities. Washington, D.C.: National Academy Press.

NRRB (National Remedy Review Board). 1998. NRB Fiscal Year 1997 Progress Report. Washington, D.C.: Environmental Protection Agency, Office of Solid Waste and Emergency Response. http://www.epa.gov/superfund/oerr/nrrb/report.htm

Probst, K. N., and M. H. McGovern. 1998. Long-Term Stewardship and the Nuclear Weapons Complex: The Challenge Ahead. Washington, D.C.: Resources for the Future.

Remediation Technologies Development Forum. 1997. Natural Attenuation of Chlorinated Solvents in Groundwater: Principles and Practices. Washington, D.C.: Environmental Protection Agency, Technology Innovation Office.

Wiedemeier, T. H., and F. H. Chapelle. 1998. Technical Guidelines for Evaluating Monitored Natural Attenuation at Naval and Marine Corps Facilities: Draft, Revision 2. Washington, D.C.: Department of the Navy.

Wiedemeier, T., J. T. Wilson, D. H. Kampbell, R. N. Miller, and J. E. Hansen. 1995. Technical Protocol for Implementing Intrinsic Remediation with Long-Term Monitoring for Natural Attenuation of Fuel Contamination Dissolved in Groundwater, Volumes 1 and 2. San Antonio, Tex.: Air Force Center for Environmental Excellence, Brooks Air Force Base.

Wiedemeier, T. H., M. A. Swanson, D. E. Moutoux, E. K. Gordon, J. T. Wilson, B. H. Wilson, J. H. Kampbell, J. E. Hansen, P. Haas, and F. H. Chapelle. 1997. Technical Protocol for Evaluating Natural Attenuation of Chlorinated Solvents in Groundwater. San Antonio, Tex.: Air Force Center for Environmental Excellence, Brooks Air Force Base.

3

Metals and Radionuclides: Technologies for Characterization, Remediation, and Containment

Many different types of inorganic contaminants are present in groundwater and soil at Department of Energy (DOE) facilities. Table 1-3 presents the results of several studies that ranked metal and radionuclide contaminants according to frequency of occurrence. In many instances, contaminants occur as mixtures of metals and radionuclides, organic complexing agents, and organic solvents. These lists of frequency of occurrence indicate the types of contaminants encountered but do not provide information about toxicity, risk, and cost-benefit of cleanup. DOE's Subsurface Contaminants Focus Area (SCFA) singled out several inorganic contaminants of concern in an informal ranking procedure based on prevalence in the weapons complex, mobility, and toxicity, as shown in Table 3-1. This chapter focuses on key contaminants from this list (U, Pu, ^{137}Cs, ^{90}Sr, Ra, Th, Tc, and Cr). The chapter does not review remediation techniques for ^{3}H because this element is generally treated using containment and natural attenuation. The chapter also does not assess remediation technologies for mercury because it is not prevalent throughout the DOE complex; it is found at Oak Ridge, primarily in sediment and surface water, not in groundwater.

FACTORS AFFECTING RISKS OF METAL AND RADIONUCLIDE CONTAMINATION

Persistence is one of the key factors considered in assessing the risk associated with a chemical in the environment. Many organic

TABLE 3-1 Inorganic Contaminants of Particular Concern at DOE Sites

Element	Medium[a]	Priority[b]	Site
Tc	GW	High	Portsmouth
Cr(VI)	GW	Medium-High	Hanford
	Soil	Medium	Hanford, Sandia National Laboratories, White Sands
U	GW	Medium-High	Fernald, Oak Ridge, Rocky Flats
	Soil	Medium	Rocky Flats, Fernald, Oak Ridge
^{137}Cs	Soil	Medium-High	Hanford, Savannah River Site
^{90}Sr	GW	Medium-High	Hanford
	Soil	Medium	Hanford
Pu	Soil	Medium	Mound, Rocky Flats, Nevada Test Site
Ra	Soil	Medium	Uranium mill tailings sites
^{3}H	GW	Medium-Low	Savannah River Site, Hanford, Brookhaven
Hg	GW	Low	Oak Ridge
	Soil	Low	Oak Ridge
Th	Soil	Low	Uranium mill tailings sites

[a]GW = groundwater
[b]Priority based on prevalence in DOE complex, mobility, and toxicity (according to a survey by the SCFA).

compounds biodegrade, reducing the potential for human and ecological exposure over the long term. Metals, on the other hand, are infinitely persistent. Radionuclides undergo natural radioactive decay that, for some compounds (such as tritium), may significantly reduce risks over relatively short time periods. However, for other radionuclides (including various isotopes of Tc, U, Pu, and Th), half-lives are very long, meaning that risks posed by the presence of these compounds will persist for a very long time. As shown in Table 3-2, half-lives for radionuclides vary quite significantly depending on the isotopes present.

The potential for humans or sensitive ecosystems to be exposed to metals and long-lived radioactive materials is strongly affected by a number of factors that must be considered in assessing these contaminants. Some metals and radioactive contaminants have more than one oxidation state, which differ in mobility and toxicity (see Box 3-1). Like organic compounds, metals and radioactive contaminants can partition into organic matter present in soils. They also can be sorbed by other soil components, including cation exchange sites and metal oxides, and they can precipitate. Because of the mul-

TABLE 3-2 Half-Lives of Radioactive Compounds Common at DOE Installations

Radionuclide	Isotope	Half-Life (years)
Tc	99	2.12×10^5
U	234	2.47×10^5
	235	7.1×10^8
	238	4.51×10^9
Cs	137	30
Sr	90	28
Pu	238	86
	239	24.4×10^3
	240	6.58×10^3
^3H		12
Th	228	1.91
	230	8.0×10^4
	232	1.41×10^{10}

SOURCE: Weast, 1980.

tiple possible associations of a metal or radioactive contaminant with a soil, determination of the total contaminant concentration is unlikely to provide sufficient information to allow valid assessments of potential risk or amenability to remediation by specific processes. Figure 3-1 summarizes the types of species in which metals may be present in the environment.

BOX 3-1
Oxidation States

Many metal and radionuclide contaminants exist in the environment in multiple forms. For example, chromium is found as Cr(VI) (+6 oxidation state) under environmental conditions known as oxidizing conditions and as Cr(III) (+3 oxidation state) under reducing conditions. Oxidizing conditions generally prevail in the absence of biodegradable organic matter and in near-surface environments. Reducing conditions generally prevail when an excess of biodegradable organic matter is present and the oxygen supply is limited.

The different oxidation states of metals and radionuclides may exhibit greatly different chemical behavior. For example, in simple, dilute, neutral aqueous solutions, the predominant form of Cr(VI) is the highly soluble, mobile oxyanion CrO_4^{2-}, while the predominant form of Cr(III) is the highly insoluble solid $Cr(OH)_3$. Metals in different oxidation states also have different risks. For example, Cr(VI) is highly toxic, whereas Cr(III) is relatively harmless to humans.

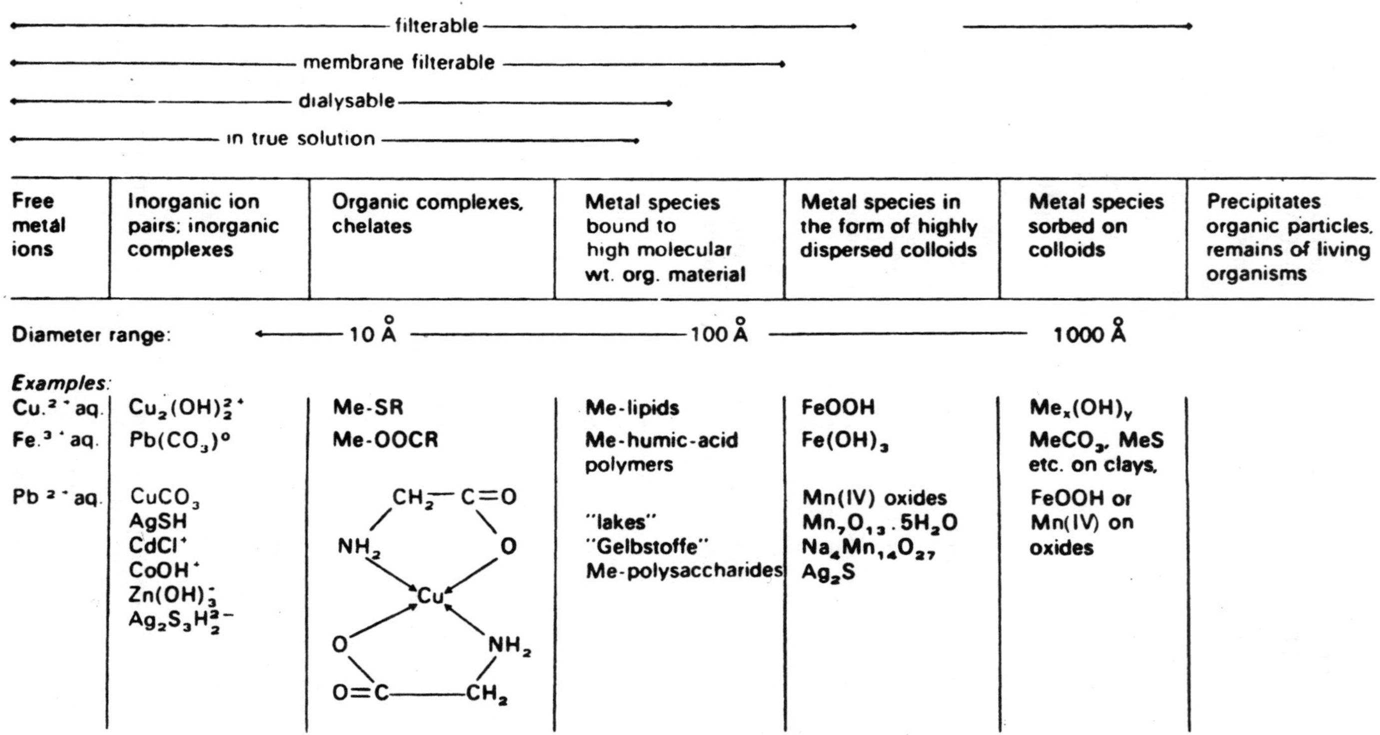

filterable

membrane filterable

dialysable

in true solution

Free metal ions	Inorganic ion pairs; inorganic complexes	Organic complexes, chelates	Metal species bound to high molecular wt. org. material	Metal species in the form of highly dispersed colloids	Metal species sorbed on colloids	Precipitates organic particles, remains of living organisms
Diameter range:	←	10 Å		100 Å	1000 Å	
Examples:						
Cu^{2+} aq.	$Cu_2(OH)_2^{2+}$	Me-SR	Me-lipids	FeOOH	$Me_x(OH)_y$	
Fe^{3+} aq.	$Pb(CO_3)^o$	Me-OOCR	Me-humic-acid polymers	$Fe(OH)_3$	$MeCO_3$, MeS etc. on clays,	
Pb^{2+} aq.	$CuCO_3$ AgSH $CdCl^+$ $CoOH^+$ $Zn(OH)_3^-$ $Ag_2S_3H_2^{2-}$	CH_2—C=O NH_2 O Cu O NH_2 O=C—CH_2	"lakes" "Gelbstoffe" Me-polysaccharides	Mn(IV) oxides $Mn_7O_{13} \cdot 5H_2O$ $Na_4Mn_{14}O_{27}$ Ag_2S	FeOOH or Mn(IV) on oxides	

FIGURE 3-1 Forms of occurrence of metal species in the environment. SOURCE: Stumm and Morgan, 1981 (*Aquatic Chemistry: An Introduction Emphasizing Chemical Equilibria in Natural Waters*, Copyright 1981 by John Wiley & Sons, Inc. Reprinted by permission of John Wiley & Sons, Inc.).

GEOCHEMICAL CHARACTERISTICS OF METAL AND RADIONUCLIDE CONTAMINANTS: EFFECTS ON TREATMENT OPTIONS

Because metal and radionuclide contaminants are generally non-degradable except by radioactive decay, treatment technologies must involve some form of mobilization or immobilization for removal or containment, respectively. Thus, solubility and the propensity for complexation in solution or sorption to surfaces are key properties to consider when evaluating treatment options. Table 3-3 summarizes these properties for the contaminants that are the focus of this study.

TABLE 3-3 Speciation of Inorganic Contaminants

A. Elements with Multiple Oxidation States

Element	Oxidizing Conditions	Reducing Conditions
Tc	Tc(VII): TcO_4^-, high solubility, very weak adsorption	Tc(IV): $TcO_2nH_2O(s)$; low solubility
Cr	Cr(VI): CrO_4^{2-}, $HCrO_4^-$, $Cr_2O_7^{2-}$ depending on total Cr concentration and pH value; high solubility, weak adsorption	Cr(III): $Cr(OH)_3(s)$; low solubility
U	U(VI): UO_2^{2+} high solubility, moderate sorption; highly soluble, weakly sorbing anionic U(VI) carbonate complexes may predominate in waters with high carbonate concentrations	U(IV): $UO_2(s)$, low solubility
Pu	Pu(VI), Pu(V), Pu(IV): Pu^{4+}, PuO_2^+, PuO_2^+ complex, redox-active aqueous chemistry with moderate solubility and moderately sorbing species	Pu(IV): $PuO_2(s)$, moderately low solubility

B. Elements with Single Oxidation States

Element	Speciation in Water
Cs	Cs(I): Cs^+, essentially no hydrolysis, moderate adsorption, no oxidation-reduction activity
Sr	Sr(II): Sr^{2+}, essentially no hydrolysis, moderate to weak adsorption, no oxidation-reduction activity
Ra	Ra(II): Ra^{2+}, essentially no hydrolysis, moderate to strong adsorption, no oxidation-reduction activity
Th	Th(IV): $Th(OH)_n^{(4-n)+}$, strong hydrolysis, moderate solubility, very strong adsorption

As shown in Table 3-3, Tc, Cr, U, and Pu exhibit multiple oxidation states, of which the reduced forms are quite insoluble in water. The oxidized forms Tc(VII) and Cr(VI) are both anions in water and generally sorb weakly to the negatively charged surfaces typically encountered in nature. The alkali and alkaline earth ions Cs^{+}, Sr^{+2}, and Ra^{+2} do not exhibit redox activity and are hard cations, which do not hydrolyze strongly, are not expected to sorb strongly to oxide surfaces, and are subject to competition in any complexation reaction by other alkali and alkaline earth ions present at much higher concentrations (e.g., Na^{+}, Ca^{+2}). Thorium(IV) does hydrolyze strongly in water and adsorbs rather strongly to oxide surfaces.

Table 3-4 summarizes treatment technologies for different classes of inorganic contaminants, and Table 3-5 summarizes technologies for different media. Box 3-2 provides a glossary of remediation technology terms. These technology options are discussed in more detail later in the chapter.

CHARACTERIZATION OF METAL AND RADIONUCLIDE CONTAMINATION

Because speciation controls the environmental transport and risks of metals and radionuclides, it is as important to characterize as the total amount (or total concentration) of the contaminant. Traditionally, concentrations of metals and radionuclides have been determined by taking samples of groundwater from monitoring wells or soil from borings to the laboratory for analysis. A variety of techniques, from computer models, to spectroscopic and electrochemical analyses, to sequential extraction methods, are available to determine speciation of metals and radionuclides in samples in a laboratory. More recently, techniques have been developed for measuring metal and radionuclide concentrations in situ, without bringing samples to the laboratory. The advantages of in situ analysis include reduction in time and cost of site characterization as well as reduction of exposure of personnel to hazardous contaminants. The following brief descriptions of laboratory and in situ techniques for characterizing metals and radionuclides are intended only as an introduction to this complex subject; the references cited with the descriptions provide technical details on carrying out these analyses.

Ex Situ Analysis for Speciation

Speciation of metals and radionuclides based on analysis of laboratory samples can be determined computationally or experimentally.

TABLE 3-4 Treatment Technology Options for Different Classes of Inorganic Contaminants

	Redox-Active Anions (Tc, Cr)	Redox-Active Actinide Cations (U, Pu)	Weakly Hydrolyzing Cations (Cs, Sr, Ra)	Strongly Hydrolyzing Cations (Th)
Solidification and Stabilization				
Pozzolanic agents	—	—	A	A
Vitrification	A	A	A	A
Chemical and Biological Reaction				
In situ redox manipulation				
Gaseous reductants	?	?	NA	NA
Liquid reductants	A	A	NA	NA
Permeable reactive barriers				
Fe^0	A	A	NA	NA
Microbiological	?	?	NA	NA
Sorption	—	—	A	A
Bioremediation	A	A	NA	NA
Biological reduction and adsorption (wetlands)	A	A	NA	NA
Separation, Mobilization, and Extraction				
Electrokinetic systems	A	A	?	?
Soil flushing and washing	A	A	A	A
Phytoremediation (macrophytes)	—	—	A	—

NOTE: A = applicable; ? = application still in an experimental stage and not yet proven; — = lack of information for a quantitative comparison; NA = not applicable.

TABLE 3-5 Technology Types Applicable to Different Contaminated Media

Context	Solidification, Stabilization	Containment	Biological and Chemical Reactions	Separation, Mobilization, and Extraction
Surface soils, sediments, and sludge	Pozzolanic agents Cement Vitrification	Capping	Phytoremediation Chemical reduction	Solvent extraction Soil washing • Acids, bases, and chelating agents • Surfactants and cosolvents
Unsaturated zone	Pozzolanic agents Cement Vitrification		Chemical reduction Microbial reduction	Electrokinetic systems Soil flushing • Acids, bases, and chelating agents • Surfactants and cosolvents
Saturated zone		Grout walls Slurry walls Sheet pile walls	Chemical reduction Permeable-reactive barriers • Fe^0 • Microbiological • Enhanced sorption • Ion exchange	Electrokinetic systems Soil flushing • Acids, bases, and chelating agents • Surfactants and cosolvents
High-concentration source areas in the saturated zone		Grout walls Slurry walls Sheet pile walls Pump-and-treat systems	Chemical reduction	Electrokinetic systems Soil flushing • Acids, bases, and chelating agents • Surfactants and cosolvents

BOX 3-2
Remediation Technologies for Inorganic Contaminants*

Stabilization-Solidification and Containment Technologies

In Situ Precipitation or Coprecipitation. A permeable reactive barrier (see definition below) that causes the precipitation of a solid (usually carbonate, hydroxide, or sulfide mineral) to maintain a toxic metal in an immobile form. Formation of solid phases is controlled primarily by pH, redox potential, and concentrations of other ions.

Pozzolanic Agents. Cement-like materials that form chemical bonds between soil particles and can form chemical bonds with inorganic contaminants, decrease permeability, and prevent access to contaminants. The most common pozzolanic agents are portland cement, fly ash, ground blast furnace slag, and cement kiln dust.

Vitrification. Melting of contaminated soil to form a glass matrix from the soil, either in place (in situ vitrification) or in a treatment unit. Nonvolatile metals and radioactive contaminants become part of the resulting glass block after cooling. Organic contaminants are either destroyed or volatilized by the extremely high temperatures. The method is generally expensive due to the large energy requirements.

Biological Reaction Technologies

Phytoremediation. Removal of contaminants from surface soil through plant uptake. Subsequent treatment of the plant biomass may be necessary.

Chemical Reaction Technologies

Enhanced Sorption. A passive-reactive barrier (see definition below) that creates zones that cause contaminant sorption, either microbiologically (biosorption) or chemically (through materials with surface complexation, ion exchange, or hydrophobic partitioning properties).

In Situ Redox Manipulation. The injection of chemical reductants into the ground

Experimental methods can be further subdivided into those that provide characteristics of the contaminant and those that provide information on specific chemical contaminants.

Computational Methods

Chemical equilibrium computer programs are useful for computing the distribution of species in samples for which total concentrations of metals and ligands (ions or molecules that can attach to metals) have been measured, provided appropriate stability constants are available (Nordstrom et al., 1979). Commonly used programs include MINTEQA2 (Allison et al., 1991) and MINEQL+ (Schecher to create reducing conditions in an aquifer, which will then lead to reduction and immobilization of certain contaminants in groundwater.

Permeable-Reactive Barriers. Permeable containment barriers that intercept contaminant plumes and remove contaminants from groundwater solution through chemical and/or biological reactions within the barrier.

Zero-Valent Iron Barrier. A passive-reactive barrier (see definition above) that creates strongly reducing conditions, resulting in hydrogen generation. Dissolved chlorinated solvents (chlorinated ethanes, ethenes, and methanes) are chemically degraded at relatively rapid rates. Some metals form relatively insoluble solids at low redox potential and can be treated with this method.

Separation, Mobilization, and Extraction Technologies

Electrokinetics. The movement of water and/or solutes through a porous medium under the influence of an applied electric field. Electromigration is the migration of ionic species through a soil matrix. The process can function in both saturated and unsaturated environments. Electroosmosis is the movement of pore water through a fine-grained matrix. This technique has long been understood as a means to control water movement in fine-grained media and is currently being investigated to remove contaminants at waste sites.

Soil Flushing. An in situ process that uses chemical amendments and fluid pumping to mobilize and recover contaminants (see also cosolvent flushing and surfactant flushing).

Soil Washing. An ex situ process in which contaminated soils are segregated and then washed with a water-based solution. Generally, soil fines have a high concentration of contaminants, while coarse materials may be sufficiently clean that contaminant concentrations are below action levels, allowing coarse materials to be disposed of separately. Once fines are separated from coarse soils, the fines may be disposed of directly or extracted.

---

* Some technologies may also be applicable to organic contaminants.

and McAvoy, 1992). Systems in which the principal ligands are inorganic and that contain relatively uncomplicated solid surfaces are most amenable to modeling with computations.

Considerable research has been conducted to describe the binding of metal ions to oxide surfaces. Applicable surface complexation models that can be used to describe this binding based on electrical double-layer models are discussed by Dzombak and Morel (1990) and Stumm (1992).

The description of metal complexation with natural organic matter (NOM; for example, humic substances) is much more complicated. NOM is an unresolvable mixture of a very large number of compounds varying in their properties, including their ability to bind metal ions. Several approaches have been proposed for modeling the way in which metals form complexes with NOM and humic substances. These include gaussian distribution models (Perdue and Lytle, 1983) and multiple discrete site models (Fish et al., 1986). Tipping (1994) has presented a model, using multiple classes of organic matter reaction sites, that is able to predict relatively accurately metal cation and proton binding to naturally occurring organic matter.

Because of the heterogeneity of soils, metals can be associated with many types of surfaces in a single soil. Recently, Radovanovic and Koelmans (1998) presented a model to predict the binding of a series of cationic metals to suspended particles in natural waters as a function of the characteristics of the aqueous and solid phases. Because similar properties control the speciation and partitioning of metals between soil and pore water, this or similar models could be applied to soil samples.

Experimental Methods

Perhaps the most fundamental physical means for determining the speciation of metals is physical separation of the dissolved and particulate phases. A number of procedures are available for this separation (Bufflap and Allen, 1995). The separation processes are subject to significant error, particularly as a result of incomplete separation of particulate and dissolved phases.

Analysis of chemical species is possible by chromatographic, spectroscopic, and electrochemical methods. The separation and quantitation of ethylenediaminetetraacetic acid (EDTA) and other complexes of metals has been achieved by ion chromatography (Hajós et al., 1996) and by capillary electrophoresis (Buergisser and Stone, 1997). Several electrochemical methods are widely used for the analysis of trace metals in natural waters and in soil solution. Among these are selective ion electrodes, anodic stripping voltammetry, and cathodic stripping voltammetry (Florence, 1989; Van den Berg, 1984). All are capable of determining submicrogram-per-liter concentrations of metals and can be used in titrations to determine the concentration of available binding sites for a metal and the strength of the complexation reaction.

Sequential extraction procedures, most commonly that of Tessier et al. (1979), frequently are used to correlate the presence of metal species in samples with observed effects, including toxicity, bioavailability (availability for uptake by living organisms), and mobility. These procedures use increasingly strong extractants to release trace metals associated with (1) exchangeable, (2) carbonate, (3) metal oxide or reducible, (4) organic and sulfide, and (5) residual mineral phases. The procedures have been criticized as being unable to provide accurate information about the associations of trace metals (Martin et al., 1987; Kheboian and Bauer, 1987; Rapin et al., 1986; Rendall et al., 1980; Sheppard and Stephenson, 1997; Tipping et al., 1985). Criticisms have focused mainly on the application of the procedures to the assessment of associations of cationic metals with specific solid phases. Sheppard and Thibault (1992) reported mixed success in using the Tessier extraction scheme for the soil litter layer, a sandy soil, and a clay subsoil that had been contaminated with Cr, Cs, Mo, Np, Pb, Tc, Th, and U. They found that the selective extraction procedure did not work well either for organic-rich soils or for anions such as TcO_4^-.

Single-extractant procedures are also widely used to estimate metal availability for uptake by plants. Among the extractants that have been used are diethylenetriaminepentaacetic acid (DTPA), EDTA, acetic acid, and the mineral acids HNO_3 and HCl (Adriano, 1986). A linear relationship between the logarithm of the concentration of the metal taken up by the plant and that extracted from soil has frequently been reported (Browne et al., 1984). The quality of predictions decreases as the soil chemistry becomes diverse. Allen and Yin (1998) suggest that the correlation failure occurs because these procedures relate to the binding phases for the metal rather than the strength of metal binding.

In Situ Chemical Analysis

One method of in situ site characterization that is increasingly being used is the incorporation of contaminant sensors into penetrometers, which are rods that are pressed into the ground. Traditionally, penetrometers were fitted with sensors to measure tip and sleeve resistance for the determination of soil stratigraphy. More recently, sensors have been incorporated into penetrometers to measure concentrations of metals, radionuclides, and other substances. These sensors include those for laser-induced breakdown spectroscopy, x-ray fluorescence (for elemental analysis), and a gamma-ray spectrometer for gamma-emitting contaminants. Each of these methods has been successfully field-tested (Ballard and Cullinane, 1997).

DOE has developed another promising technology for radiological characterization of soil surfaces, particularly for use in association with excavation activities. The technology, known as the digface sensor, depends on an appropriate sensor, a precise *x-y-z* positioning system, a method to move the sensor systematically over the area to be investigated, and a data reduction and display device (Josten et al., 1995). Advantages include quality real-time data for better decision making, a potential to reduce the amount of material requiring excavation, reduction in hazard to personnel, and reduction in the risk that contaminants will inadvertently be left in place. The system has been successfully applied for ^{232}Th, ^{227}Ac, ^{137}Cs, and ^{238}Pu contamination in field tests. (See Chapter 5 for more information on this technology.)

PHYSICAL BARRIERS FOR CONTAINING CONTAMINANTS

Barrier systems are among the most widely used technologies for managing contaminated sites. A wide variety of designs has been developed to meet particular needs. Physical barriers can be grouped into three categories: (1) vertical barriers, (2) surface caps, and (3) emplaced horizontal barriers (bottoms).

Vertical barriers can provide rapid and significant risk reduction by isolating the contaminant source from the flowing groundwater. They also can provide opportunities for enhanced remediation by controlling groundwater hydraulics and/or allowing chemical treatment of the aquifer that would not be possible without physical containment. In the context of many DOE-related groundwater problems, another important characteristic of vertical barriers is their potential to stabilize contamination over periods of years to decades. Such contaminant stabilization allows time for chemical degradation, radioactive decay, or the development of improved remediation technologies.

Surface caps, such as those used on modern landfills, are also widely used at many DOE facilities to control infiltration and water movement through contaminated soils. Because of their location at the surface, caps generally have a more sophisticated layered structure than vertical barriers and can be instrumented much more easily with water collection systems and sensors. Their role in groundwater contaminant transport is limited to reducing leaching from the vadose zone to the groundwater. As a consequence, surface caps are not discussed further in this chapter.

Like surface caps, horizontal barriers are widely used beneath modern municipal, hazardous waste, and DOE landfills. Emplacement of horizontal barriers beneath existing uncontained sources of groundwater contamination is likely to become more common. These constructed "bottoms" are potentially quite important in the context of DOE's dense nonaqueous-phase liquid (DNAPL) and metal contamination problems because they may be able to minimize downward migration of the contaminants.

For additional detail on vertical and bottom barriers beyond that presented below, see Rumer and Mitchell (1995) and DOE (1997).

Vertical Barriers

In conceptually simple form, a vertical barrier can be used to completely surround a source of groundwater contamination with an essentially impermeable wall. Ideally, this wall can be physically connected to a naturally occurring horizontal barrier (for example, an aquifer confining unit) and result in isolation of the contamination source from the groundwater. In practice, all barriers and confining units have some level of permeability, and as a consequence, water can move into or out of the contained aquifer. This fact, and the desire to eliminate any contaminant movement from the source, frequently means that a small-scale pump-and-treat system is coupled with the barrier to maintain a constant inward hydraulic gradient across all faces of the barrier. If containment is not complete (for example, if there is no confining unit), pumping at significantly higher rates may be necessary to maintain inward gradients.

Another hydraulic aspect of vertical barriers is that the presence of the barrier may affect the surrounding groundwater flow. For example, if a site is completely surrounded by a vertical barrier, groundwater will "mound up" at the upgradient edge of the barrier. As a consequence, groundwater hydraulically upgradient of the site will be deflected around the site. Although this may not affect the site directly, it could affect adjacent sites.

Partial barriers that do not completely surround the contaminant source area also can be used for containment. Partial barriers are used primarily for plume capture to prevent off-site migration. As with fully contained systems, the presence of the barrier will cause mounding of the water table and lateral and potentially downward diversion of the plume. As a consequence, barriers constructed perpendicular to groundwater flow will have to extend upgradient for some distance to ensure containment. In addition, changes in regional flow direction can cause the contaminant plume to shift and miss the barrier. For these reasons, extending the upgradient portions of the barrier (the "wings") to the point where they are cross-gradient from the source may be necessary. Thus, modeling groundwater flow and contaminant transport in relation to barriers is a critical component of barrier design (Rabideau et al., 1996; Russel and Rabideau, 1997; Smyth et al., 1997a).

Geologic Conditions for Barrier Emplacement

Geologic conditions must be suitable for emplacement of vertical barriers, or conversely, the installation method must be compatible with geologic conditions at the site. From a geologic perspective, the important parameters will include aquifer permeability, heterogeneity, the presence of bedrock or large cobbles, the presence of an aquitard, and the depth to the bottom of the contaminated zone.

All techniques used for emplacement of vertical barriers have overall depth limitations, even in geologically favorable conditions. For example, excavators have an operational depth limit of approximately 15 m (50 ft), and sheet pile can generally be driven only to depths of 30 to 45 m (100-150 ft). For other techniques such as deep-soil mixing or jet grouting, the necessity to interlock barrier panels at depth may ultimately limit the depth to which they can be applied. At many sites, aquifer characteristics may limit the applicability of many or all barrier techniques. For example, fractured bedrock aquifers are not well suited for most barrier technologies. Similarly, installing barriers in aquifers consisting of large cobbles may be difficult.

Effects of Contaminant Properties and Site Conditions

In addition to hydrologic and geologic conditions, the success of vertical barriers may depend on a number of other processes related to contaminated properties and site conditions. These can include physical processes, such as molecular diffusion, and chemical processes, such as sorption, ion exchange, dissolution-precipitation, and oxidation-reduction. In some cases, the presence of the waste may cause geochemical changes that affect barrier integrity (for example, shrinking and cracking of the barrier due to geochemical weathering or the presence of solvents or destruction of the barrier caused by the presence of strong acids, bases, or solvents). As a consequence, understanding possible interactions among the barrier, the geochemistry of the subsurface, and the contaminants is essential.

Installation Methods

The choice of installation methods for vertical, low-permeability barriers depends on a number of factors. Table 3-6 summarizes five different categories of installation procedures.

Trenching Trenching is the excavation of native materials and their replacement with lower-permeability media such as clayey soils (see Figure 3-2). This procedure is usually accomplished with exca-

TABLE 3-6 Installation Procedures for Vertical Barriers

Wall Type	Width, m (ft)	Depth, m (ft)	Unit Cost, $\$/m^2$ ($\$/ft^2$)	Production Rate, m^2/10 h (ft^2/10h)
Soil bentonite trench	0.6-0.9 (2-3)	24 (80)	22-86 (2-8)	230-1,400 (2,500-15,000)
Cement bentonite trench	0.6-0.9 (2-3)	24 (80)	54-190 (5-18)	93-740 (1,000-8,000)
Deep soil mixing (DSM)	0.76 (2.5)	27 (90)	65-160 (6-15)	93-280 (1,000-8,000)
DSM structural	0.76 (2.5)	27 (90)	160-320 (15-30)	93-280 (1,000-3,000)
Jet grouting	0.46-0.91 (1.5-3)	61 (200)	320-860 (30-80)	28-230 (300-2,500)
Cryogenic[a]	0.30-3. (1-10)	61+ (200+)	650 (60)	—
Sheet pile	<0.03 (<0.1)	30 (100)	160-430 (15-40)	93-280 (1,000-3,000)

[a] From Peterson et al., 1996.

SOURCE: Adapted from Filz et al., 1996.

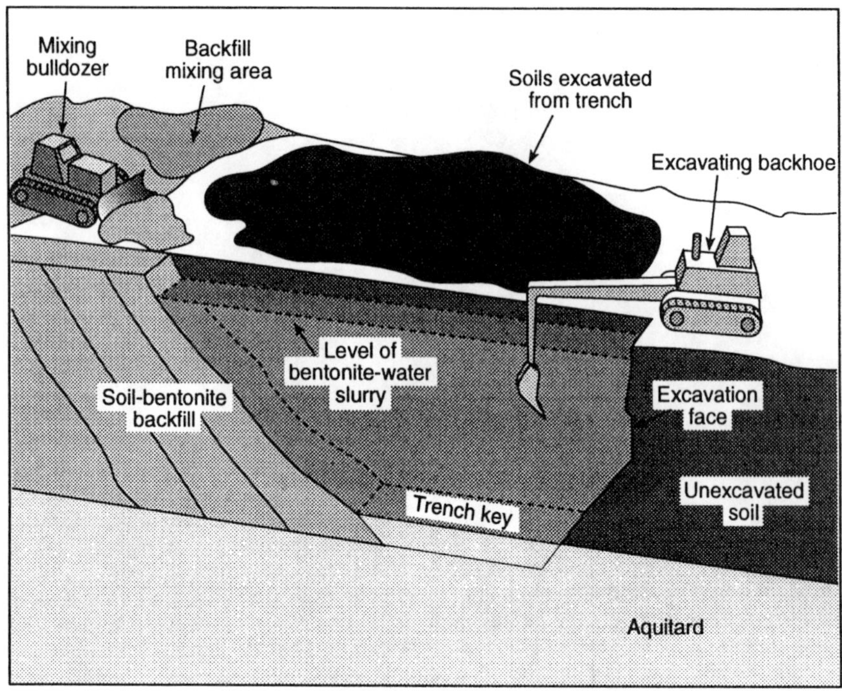

FIGURE 3-2 Installation of a soil-bentonite cutoff wall. SOURCE: Rumer and Ryan, 1995.

vators or shovels, depending on the depth and materials. Below the water table the trenches generally have to be kept open until backfilled. In some applications, excavation and backfilling are carried out almost concurrently. In others, the trench is held open by mechanical supports or by viscous fluids such as guar gum.

Trenching is best suited to large sites in which the contaminant zone can be surrounded while a minimum amount of contaminated soil is excavated and at which complete treatment of the contaminated materials would be prohibitively expensive. Like other physical containment methods, trenched barriers have limited application at depths greater than 30 m (100 ft). Unfortunately, this depth limitation restricts the use of trenched barriers as a treatment option at a number of DOE facilities.

Pressurized Injection Pressurized injection involves injecting grout into the subsurface under pressure. In general, pressurized injection barriers are constructed by intersecting short sections of grout wall (e.g., columns or sheets). The final composition of the barrier is a mixture of the native materials and the injected material. Injection can occur at very high pressure, in which the soil structure is disrupted and the soil is mixed with the injected material (jet grouting). Injection can also occur at lower pressure using low-viscosity materials that move into the existing soil structure (permeation grouting). Pressurized injection is applicable when conventional trenching is not practical, either because of space constraints or because excavation of contaminated soils cannot be accomplished. An advantage of pressurized injection relative to trenches is that in general, a substantially smaller volume of soil has to be excavated. However, constructing an intact pressurized injection barrier is generally much more difficult than constructing a conventional trench. To date, there have been relatively few thorough examinations of the permeability of pressure-injected barriers. Based on work to date it appears that the permeabilities of pressurized injection barriers are not as low as those of conventional trenched barriers because of the difficulty in ensuring complete connection of the barrier sections at depth.[1]

Driven Rigid Barriers (e.g., sheet pile) To date, most environmental applications of driven sheet pile have occurred in research settings, where these barriers have proven useful for controlled field studies (Smyth et al., 1997b). More recently, the use of driven sheet pile is increasing in "funnel-and-gate" applications of permeable barriers, in which the sheet pile directs contaminated groundwater into a wall section containing media designed to react with the contaminants. Sheet pile used for these purposes is specifically designed and includes grout or gaskets to seal joints and minimize leaks.

A number of potentially significant limitations to the use of driven sheet piles remain. Their use is limited to locations at which the sheets can be driven (e.g., areas with no cobbles or boulders) and to depths of approximately 30 m (100 ft). In some cases driving sheet piles has caused land subsidence and foundation damage. The process of driving may also mobilize contaminants and provide pathways for vertical contaminant transport. In addition, there is some concern that corrosion could limit the lifetime of steel sheet pile,

---

[1] As an example of the effect of minor breaks in the overall hydraulic conductivity (K) of a barrier, consider an otherwise impermeable barrier ($K = 0$) in a permeable aquifer ($K > 10^{-2}$ cm/sec). An opening of only 1 m^2 per 10,000 m^2 of barrier will result in effective conductivities greater than 10^{-6} cm/sec.

although corrosion is generally significant only under oxidizing conditions and subsurface contamination frequently creates reducing conditions in the aquifer. Finally, sheet pile is a relatively expensive barrier technology at present. Nonetheless, it has a number of potentially important advantages. First, installation is rapid, and soil excavation is not required. Perhaps equally as important, sheet pile can be removed when it is no longer needed. In this context, sheet pile couples well with aggressive in situ treatment technologies, such as chemical flooding systems (discussed later in this chapter).

Deep-Soil Mixing Deep-soil mixing involves mixing contaminated soil with chemical agents to treat contaminants directly and to improve treatment by homogenization (i.e., removal of heterogeneities) (Siegrist et al., 1995; Korte et al., 1997) (see Figure 3-3). Its use for reduction of the overall permeability of the soil by the addition of cements or grouts is uncommon (Filz et al., 1996).

Cryogenic Barriers Cryogenic barriers are formed by freezing the soil to prevent transport of water and contaminants. As with driven sheet pile, these barriers were initially used in the construction industry, and their use in environmental applications is more recent. Cryogenic barriers have a number of desirable traits, which include the following: (1) they are self-healing; (2) their removal can be accomplished by letting the barriers thaw; and (3) they couple well with directional drilling techniques, which can allow barriers to be installed as "bottoms." However, cryogenic barriers are not without shortcomings. Primary among these is that they require ongoing operation and maintenance. In addition, they can be used only in water-saturated soils (Dash et al., 1997; Lesmes et al., 1997; Peters, 1994; Peterson et al., 1996; Williams et al., 1997).

Installation Materials

A wide variety of materials are currently being used for vertical barriers. Their applicability depends on installation techniques, hydrogeologic conditions (including soil type and especially depth), compatibility with relevant contaminants, the time frame over which the material is to be used as a barrier, and cost. Barrier materials and some of their characteristics are listed in Table 3-7, along with selected references.

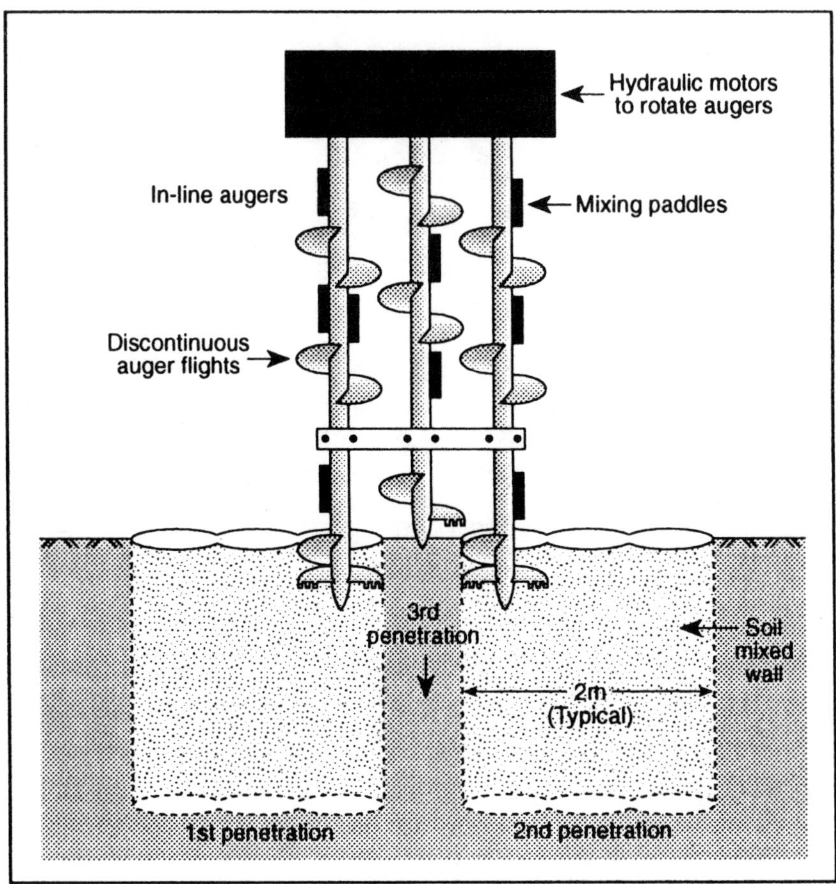

FIGURE 3-3 Deep-soil mixing for installation of a vertical barrier. SOURCE: Filz et al., 1996.

Bottom Barriers

Bottom barriers, emplaced beneath existing in situ contaminants, have a number of features in common with vertical barriers, including the fact that (1) they can be used either in the vadose zone or beneath the water table, (2) their primary function is to minimize groundwater flow using low-permeability materials, (3) they can be constructed of similar materials, and (4) they can be emplaced using some similar technologies. However, the range of options for constructing bottom barriers is significantly less than that for vertical barriers (Peterson et al., 1996).

TABLE 3-7 Barrier Materials

Material	Reference[a]	Metal Compatibility	DNAPL Compatibility	Lifetime (years)	Hydraulic Conductivity (cm/sec)	Cost, $/m² ($/ft²)
Cement	1	Good	Good	>25	10^{-8} – 10^{-9}	54-190 (5-18)
Bentonite	1	Good	Good	>25	$<10^{-7}$	22-86 (2-8)
Cement replacement materials	1	Good	Good	>25	10^{-6} - 10^{-9}	22-160 (2-15)
Sheet piles	2	Good	Good	>25	10^{-9}	160-430 (15-40)
Geomembranes	3	Good	Good	>25	10^{-9}	54-270 (5-25)
Sodium silicate	4	Good	Fair	10-20	10^{-5}	130 (12)
Acrylate gel	5,6	Good	Fair	10-20	10^{-7} - 10^{-9}	230 (21)
Colloidal silica	6	Good	Good	>25	10^{-8}	54-320 (5-30)
Iron hydroxides	6	Good	Good	>25	10^{-7}	54-160 (5-15)
Mantan wax	6	Good	Fair	>25	10^{-4} - 10^{-7}	320 (30)
Sulfur polymer cement	6	Good	Fair	>25	10^{-10}	180 (17)
Epoxy	6	Good	Good	>25	10^{-10}	480 (45)
Polysiloxane	6	Good	Good	>25	10^{-10}	54-1,600 (30-150)
Furan	6	Good	Good	>25	10^{-8} - 10^{-10}	800 (75)
Polyester styrene	6	Good	Fair	>25	10^{-10}	970 (90)
Vinylester styrene	6	Good	Good	>25	10^{-10}	1,100 (100)
Acrylic	6	Good	Good	>25	10^{-9} - 10^{-11}	1,900 (180)
Cryogenic	7	Good	Good	>25	10^{-5} - 10^{-9}	160-430 (15-40)

[a]1 = Evans et al., 1996; 2 = McMahon et al., 1996; 3 = Koerner et al., 1996; 4 = Whang et al., 1996; 5 = Peterson et al., 1996; 6 = Persoff et al., 1994.

Installation Methods

The most straightforward approach to installing bottom barriers is angled drilling. One such method involves drilling a set of angled holes down either side of a contaminated site to form a V-shaped trough beneath it. The angled holes can be used to introduce low-permeability materials (e.g., using jet grouting) or as access holes for a cryogenic barrier. Directional drilling can also be used, in which case a continuous hole under the contaminated zone (and potentially up to the ground surface on the other side) could be emplaced (see Figure 3-4).

Materials

The installation method generally determines which materials can be used to construct bottom barriers. For example, materials suitable for jet grouting will be appropriate for installations using angled drilling. However, at this time there is little experience either within DOE or in the private sector in constructing bottom barriers by using directional drilling. Given the trend in recent years toward risk reduction by containment, significant advances likely will be made in bottom barrier installation by directional drilling in the next few years. Installation of cryogenic bottom barriers, using either angled or directional drilling, is also likely to increase in the next five years. Nonetheless, all of the bottom barrier technologies are likely to remain expensive for the foreseeable future. In addition, verification of bottom barrier performance is likely to remain a challenging task (Peterson et al., 1996).

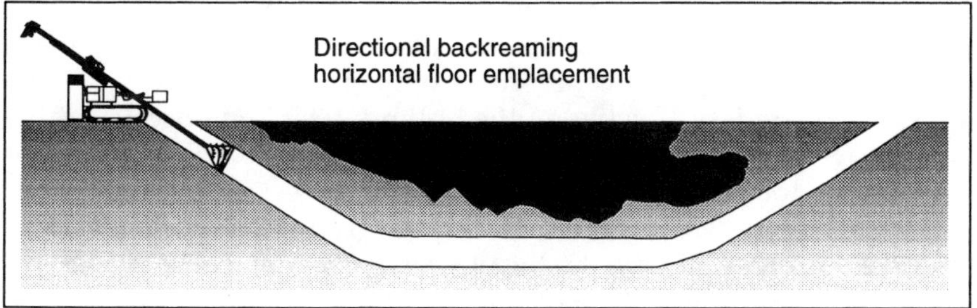

FIGURE 3-4 Creation of a bottom barrier using directional drilling. SOURCE: Peterson et al., 1996.

Verifying Barrier Performance

Performance verification is a critical component in the use of barriers, and planning for verification should be an integral part of decisions regarding barrier design and materials. The method(s) used for verification will depend on the hydrogeologic setting (e.g., above or below the water table) and the risk posed by the contaminants. Four approaches to barrier verification are (1) hydraulic tests, (2) tracer tests, (3) emplaced sensors, and (4) geophysical methods.

Hydraulic Tests

Hydraulic tests are useful for characterizing the large-scale features of a barrier (Snyder et al., 1997). For example, if a barrier completely encloses a portion of an aquifer and is closed at the bottom, the presence of relatively small defects in the barrier can be deduced from pumping tests. However, if the barrier only partially encloses a site, then hydraulic tests are relatively insensitive to small leaks. Hydraulic tests also frequently provide very little detail about the locations of leaks.

Tracer Tests

Tracer tests can pinpoint leaks in barriers, but they require installation of numerous monitoring points around the barrier (Williams et al., 1997). Using tracer tests for barrier verification has proven very feasible in the vadose zone, where gas-phase tracers can be used and diffusion coefficients are relatively high and isotropic. The ability of detection systems to identify accurately the locations and extents of leaks requires a clear understanding of gas-phase diffusion within the soils surrounding the barrier. To date, such tests have been applied primarily in relatively dry soils, where the diffusion coefficients are large. They have not yet been demonstrated in complex geologic systems where the diffusion coefficient may vary spatially and/or temporally by several orders of magnitude.

Tracers are less effective in the groundwater zone because diffusion is much slower than in the gas phase and because advection frequently dominates, so the probability of detecting the tracer with a monitoring network is much lower. In groundwater, the contaminants themselves may be the best tracers. The detection of contaminants in monitoring wells indicates that the barrier has leaked.

Emplaced Sensors

A wide range of physical and chemical sensors has been developed in the last 10 years (Borns, 1997; Inyang et al., 1996). These sensors can be deployed in a variety of ways, either within a barrier or adjacent to it. Depending on the hydrogeologic and chemical setting, these sensors may directly indicate the presence of water (e.g., in the vadose zone), or they may be selective for specific chemical contaminants. Once again, in the vadose zone, sensors to detect gas-phase transport are likely to be quite successful. In the groundwater zone, the difficulty is that given very limited dispersion, effectively covering potentially hundreds of square meters of barrier would require hundreds if not thousands of sensors. At present this is not a practical alternative.

Geophysical Methods

Geophysical methods have great potential for verifying barriers. However, although these methods have been applied effectively in some cases, a significant number of less successful applications also have occurred. In some cases, problems were due to inadequate resolution of the instruments. In others, the hydrogeologic conditions were inappropriate for a given geophysical technique. A number of potentially useful geophysical techniques for barrier verification are listed in Table 3-8, along with general comments about their application. In general, application of these techniques for barrier verification will

TABLE 3-8 Geophysical Methods for Barrier Verification

Geophysical Method	Comments	References
Ground-penetrating radar	Resolution 0.5-1 m with cross-borehole, tomographic method	Pellerin, 1997 Davis and Annan, 1989 Lesmes et al., 1997
Electromagnetics	Resolution >0.5 m in vicinity of borehole	Pellerin, 1997
Electrical resistivity	Resolution depends on electrode spacing, typically 0.3-2 m; tomographic method	Pellerin, 1997 Daily and Ramirez, 1997
Seismic	Tomographic method	Pellerin, 1997 Steeples and Miller, 1993

require the collection of detailed, three-dimensional data sets and the tomographic imaging of these data using numerical methods.

TECHNOLOGIES FOR IMMOBILIZING METALS AND RADIONUCLIDES

Immobilization of metal and radionuclide contaminants by physical, chemical, or biologically mediated binding of the contaminant in some way within the soil matrix is becoming an increasingly common approach to site remediation. The broad categories of immobilization technologies discussed in this section include in situ vitrification, solidification and stabilization, permeable reactive barriers, in situ redox manipulation, and bioremediation.

In Situ Vitrification

Description

In situ vitrification (ISV) is an immobilization and destruction technology designed to treat soils and other similar media contaminated with heavy metals, organic compounds, and radionuclides. Soils are heated and melted by applying an alternating electrical current between electrodes placed in the ground (see Figure 3-5). Paths of graphite and glass frit are placed in the soil between the electrodes to aid in the start-up of the ISV process. The temperature of the molten soil may exceed 1700°C. At these temperatures, organic compounds either volatilize and are captured in a hood or are destroyed. Upon

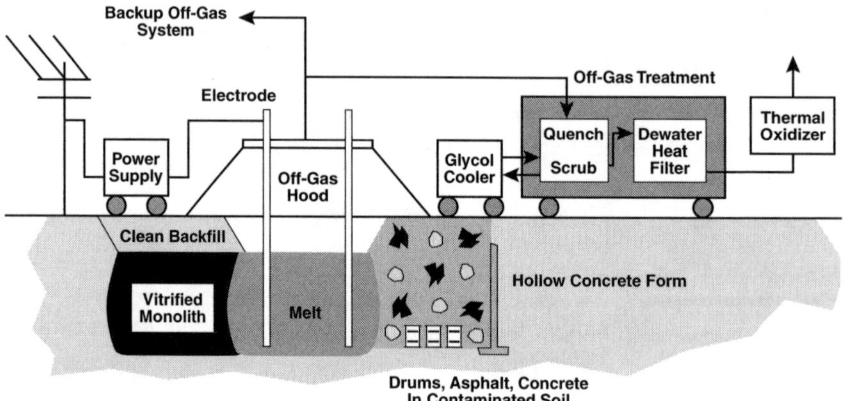

FIGURE 3-5 Schematic of an ISV System. SOURCE: GeoSafe, Inc.

cooling, the molten soils become an impermeable glass or crystalline solid. The glass or crystalline solid is very leach resistant compared to the original untreated soil; metals and radionuclides may be chemically bound to the matrix or physically entrapped. As a result, the vitrified material can remain without significant risk to human health or the environment. Secondary residuals generated by ISV, which typically require treatment, include air emissions, scrubber liquids, carbon filters, and used hood panels.

Physical and Chemical Principles

The development of in situ vitrification was based on principles of ex situ vitrification processes, which are well established. Soils containing sufficient concentrations of conductive cations are slightly conductive to electricity, so electric currents can pass through. The relatively high resistivity of soil requires high voltages to achieve flow of the electrical current. Ultimately, passing a current through the soil creates large amounts of heat, which melts the soil.

When raised to temperatures exceeding its melting point, soil forms a liquid, or melt, that upon cooling forms either a glass or a crystalline material. Glasses and crystalline materials are highly impermeable and have extremely small surface areas compared to untreated soils. As a result, vitrified soils leach poorly, and metals and/or radionuclides contained within the vitrified soil are very immobile. The heating process may also cause some metals and/or radionuclides to bond chemically to the vitrified soil matrix, further reducing mobility.

The high temperatures achieved during vitrification either pyrolize or volatilize the organic compounds that may be present along with metals or radionuclides (see Chapter 4). The increased vapor pressures caused by the heating and the creation of a low-pressure zone in the overlying hood cause the organic vapors to migrate to the hood, where they are captured and treated.

Application

In the vitrification system developed by Battelle Pacific Northwest Laboratories and licensed to Geosafe, Inc., four graphite electrodes are inserted to a shallow depth in a square pattern in the soil to be treated. A pattern of electrically conductive graphite and glass frit is placed on the soil to complete an electrical circuit between the electrodes. A large electrical source (usually trailer mounted) is applied to the electrodes. The electrical resistance results in heating of the glass frit and graphite path and the soils near the path. As heating continues, the melting of soils continues outward from the graphite and glass frit path because molten soils are electrically conductive and also transfer heat to adjacent soils. As the melting continues downward, the electrodes are moved deeper into the soil. The melting is continued until the targeted depth is reached.

Contaminant vapors formed during ISV can be captured by placing a hood over the melt area and applying a partial vacuum to the hood. Off-gas treatment is required. Monitoring of off-gases (stack emissions) is conducted for organic compounds present in the untreated matrix: oxygen, carbon dioxide, carbon monoxide, particles, metals, and hydrogen chloride. These analyses are compared to discharge permit requirements and system performance criteria. The quenched and scrubbed water is analyzed for organics, pH, and metals to determine if discharge requirements are being met.

Once the targeted soils have melted, the electric current is turned off, and the soils are allowed to cool. The electrodes are cut off at the surface and allowed to become part of the melt. Over an extended period the soil solidifies into a glass and/or crystalline monolith. Following adequate cooling of the surface, typically 24 hours or longer, clean soil is placed on the surface of the treated area to make up for the subsidence created by the soil consolidation that results from melting. Once cooling occurs, the treated matrix is sampled for constituents of concern. Analyses are made for total constituents, and the toxicity characteristic leaching procedure (TCLP) is carried out, if required, to determine the toxicity of leachate from the monolith.

Modifications of ISV include various staged in situ alternatives in which excavated material is placed in a subsurface zone for treatment, either in a single layer or in multiple layers. In the latter, material can be treated below grade, and additional material can be added for further treatment. The reverse can be accomplished by treating the upper layer of material, removing the glassified matrix, and then treating the next layer. Methods also have been developed to address soils with insufficient cation content.

Prior to implementing ISV, treatability tests are conducted on 45- to 90-kg (100-to 200-lb) soil samples to determine heat requirements and vitrified product properties. Design considerations include the lateral dimensions, depth, and composition of the matrix to be treated. Moisture is also an important consideration.

Current technology can be implemented to a maximum depth of 6 m (20 ft). If intact steel drums containing organic liquids are present, pretreatment is required to rupture the drums. If the amounts of organic compounds are excessive, heat may damage the equipment. Appropriate modifications for soils with high organic content include lower melt rates or dilution of the matrix.

ISV is applicable to sludges, sediments, and soils. ISV has been demonstrated to immobilize heavy metals and radionuclides and to remove and/or destroy volatile and semivolatile organic compounds, polychlorinated biphenyls (PCBs), dioxins or furans, pesticides, and munitions (discussed in more detail in Chapter 4). ISV can treat soils contaminated with large amounts of metals, although molten metal may sink to the bottom of the melt. Buried steel drums are not a problem unless they contain liquids and have maintained their structural and sealing integrity. These drums will eventually fail, releasing vapors to the melt in a potentially disruptive fashion. Pretreatment is required for intact steel drums containing organic liquids. ISV can tolerate waste and debris within the treatment zone. Organic debris is destroyed primarily by pyrolysis. Inorganic debris is typically incorporated into the melt and vitrified product. Examples of debris that have been present in ISV-processed soils include wood, vegetation, plastic, rubber, cardboard, asphalt, oils, and construction materials.

Performance

Table 3-9 lists several sites at which ISV has been used to treat radionuclides and metals.

TABLE 3-9 Applications of In Situ Vitrification

Site, Location	Constituents	Comments
Parsons Chemicals Works, Grand Ledge, Mich.	Pesticides heavy metals	Excavated materials 14,800 tons Met all cleanup standards
Oak Ridge National Laboratory	Cesium-137	Melt exploded; test terminated
Maralinga, Australia	Radionuclides heavy metals	Demonstration program Inorganic debris ~50% volume reduction >99.9998% retention of U and P
Hanford Site	9 radionuclides, 13 metals	20 pilot-scale tests; 6 large-scale tests
Ube City, Japan	Organics, PCBs, heavy metals	Tests with soil, mortar, asphalt, and drums

SOURCE: Geosafe, Inc., personal communication, 1998.

Limitations

ISV has several limitations. One limitation is that the soil organic content must be less than about 7 to 10 percent under normal operations. ISV can accommodate somewhat higher organic content with a reduced power level, but this slows the rate of treatment. Limitations on organic content are a result of the release of organic vapors, which occurs primarily through the dry soils along the edge of the melt. If, as a result of high organic content, organic vapors pass through the melt, the heat removal capacity may be exceeded. Sputtering and splattering of molten material may result, damaging the vapor collection hood.

The presence of rapidly recharging water within the treatment matrix, such as would occur within permeable aquifers, can also cause problems with ISV. Prior to melting of the matrix, water vaporizes and escapes upward along the outside of the melt. Condensation of moisture a short distance (typically one-third of a meter or so) outside the melt can create a saturated soil barrier and temporarily trap organic vapors. Sudden releases of steam under pressure can occur, causing overpressurization of the above-ground system. This may have been the cause of an accident that occurred during ISV implementation at the Oak Ridge, Tennessee, DOE facility, in which 20 tons of molten product erupted from the subsurface and damaged the off-gas hood (Spalding et al., 1997). No personnel were injured, but the project was terminated as a result of the accident.

For ISV to be effective, the matrix to be vitrified must contain sufficient conductive cations (sodium, lithium, magnesium, etc.) for the molten mass to be adequately conductive. Additionally, the matrix should contain adequate amounts of glass-forming elements such as silicon and aluminum (seldom a problem in soils).

ISV is not appropriate without some form of pretreatment or alternative approach (such as excavation of the upper several feet of soil or addition of alumina or silicate) under the following conditions:

- depth greater than about 6 m (20 ft);
- excessive moisture levels or high moisture recharge rates, especially if volatile organic compounds are present;
- presence of operational utility trenches within the treatment zone;
- presence of intact steel drums containing organic liquids;
- a matrix composition that results in an excessively high melting temperature or that will not form a glass and/or crystalline product upon melting and cooling;

- a soil organic content exceeding safe levels (7 to 10 percent by weight);
- a metal content exceeding 15 percent by weight (which requires an electrode feeding procedure);
- inorganic debris exceeding 20 percent by weight; or
- inadequate surface area to set up above-ground components, including a crane for placing the vapor recovery hood.

Advantages

Although it has a number of limitations, ISV also has some unique advantages. A significant advantage of ISV is that it can treat complex matrices containing mixtures of contaminant types in a single step, a capability that few technologies share. Another significant advantage for DOE sites is that treatment can occur without bringing radioactive materials to the surface, which can help reduce exposure risks and potential transportation problems. An additional possible advantage of ISV is that the cooled vitrified mass can serve as a foundation for various types of construction, allowing for a wide range of uses of the area where treatment occurred.

Solidification and Stabilization

Description

Solidification and stabilization processes are designed to reduce the mobility of contaminants by reducing the contaminant solubility or the permeability of the medium (NRC, 1997). Solidification is the formation of a stabilized mass in which the contaminants are physically bound or contained. In stabilization, chemical reactions are induced between the stabilizing agent and the contaminant to reduce mobility. Both ex situ and in situ methods are available. Ex situ processes are among the most mature technologies, and excavated soils are frequently treated prior to disposal. Solidification and stabilization procedures have been described by Smith et al. (1995) and EPA (1994, 1997c).

Physical and Chemical Principles and Application

Principal stabilization materials are portland-type cements, pozzolanic materials, and polymers:

• Portland cements typically consist of calcium silicates, aluminosilicates, aluminoferrites, and sulfates. Metals are immobilized in cement-type binders as hydroxides or other stable solids.

• Pozzolans are very small spherical fly ash particles formed in the combustion of coal, in lime and cement kilns, and in other combustion processes. Those that are high in silica content have cement-like properties when mixed with water.

• Polymeric compounds can be used to bind metal and radionuclides by microencapsulation. Materials that have been investigated for this purpose include bitumen, which is the least expensive, as well as polyethylene and other polyolefins, paraffins, waxes, and sulfur cement. DOE has used polyethylene encapsulation to treat a number of radionuclides (including cesium, strontium, and cobalt) and toxic metals (including chromium, lead, and cadmium).

Introduction of chemical reagents in stabilization processes can cause in situ chemical modification. For example, hydrogen sulfide has been used to precipitate metals (IAEA, 1997). However, such procedures lack good operational control, and the process efficacy is not known. A significant effort is currently being directed toward the application of low-cost amendments to soils to immobilize lead. Many of these processes involve the formation of secondary minerals, such as metal phosphates. The monolith is left onsite or landfilled. The stability of these materials has not been subjected to long-term testing, but it is expected to be significantly better than that of hydroxide precipitates.

Performance

Geo-Con, Inc., reportedly has used in situ solidification and stabilization at dozens of sites in the United States (EPA, 1995a). Projects have included construction of a 20-m-deep soil-bentoite wall to contain groundwater contamination in a former waste pond and shallow soil mixing and stabilization of 82,000 yd^3 of contaminated soil at a former manufactured gas plant site. In a demonstration of the process conducted under the SITE (Superfund Innovative Technology Evaluation) program, the permeability of the treated soil decreased from 10^{-2} cm/sec to between 10^{-6} and 10^{-7} cm/sec. Polychlorinated biphenyl immobilization appeared likely as a result of mostly undetectable PCB concentrations in leaching tests on the treated soil, although this conclusion could not be confirmed because of low PCB concentrations in the untreated soil. However, data collected during this demonstration were insufficient to evaluate the effectiveness of the process in immobilizing metals.

Limitations

The success of in situ solidification or stabilization depends on the ability to mix the stabilizing agent with the soil. However, ensuring that sufficient mixing has occurred is difficult. Soils with high clay content or with large amounts of debris are not suitable for this treatment method. The process is not effective for anionic species such as As(III), As(V), and Cr(VI) because these remain mobile after treatment (Evanko and Dzombak, 1998). Ex situ mixing may cause the release of organic vapors.

Advantages

Solidification or stabilization processes are broadly applicable to a wide variety of metals and to wastes that contain mixtures of metals and some types of organic compounds (EPA, 1997c).

Permeable Reactive Barriers

Description

A permeable reactive barrier is a passive in situ treatment zone of reactive material that immobilizes metal or radionuclide contaminants as groundwater flows through it (Vidic and Pohland, 1996; EPA, 1998; Schultz and Landis, 1998). In this type of system, a permeable treatment wall is installed or created across the flow path of a contaminant plume (see Figure 3-6). Sorption or precipitation reactions occurring within the barrier remove metals and radionuclides from the groundwater, immobilizing the contaminants.

Physical and Chemical Principles

Sorption and precipitation reactions for treating metals and radionuclides in reactive barriers can be brought about through various physical, chemical, and biological processes. For example, inorganic contaminants can be sorbed to zeolites, hydrous ferric oxide, peat, silica, and polymer gels; reduced and precipitated by Fe^0, ferrous hydroxide, H_2S, or dithionite; or precipitated by lime or limestone. Biologically active zones can reduce and precipitate, as well as sorb, inorganic contaminants.

Because metals do not degrade and because most radioactive contaminants do not decay on the time scale of interest, the reversibility of the immobilization reaction must be closely scrutinized. If the immobilization reaction is reversible on the time scale of interest, the

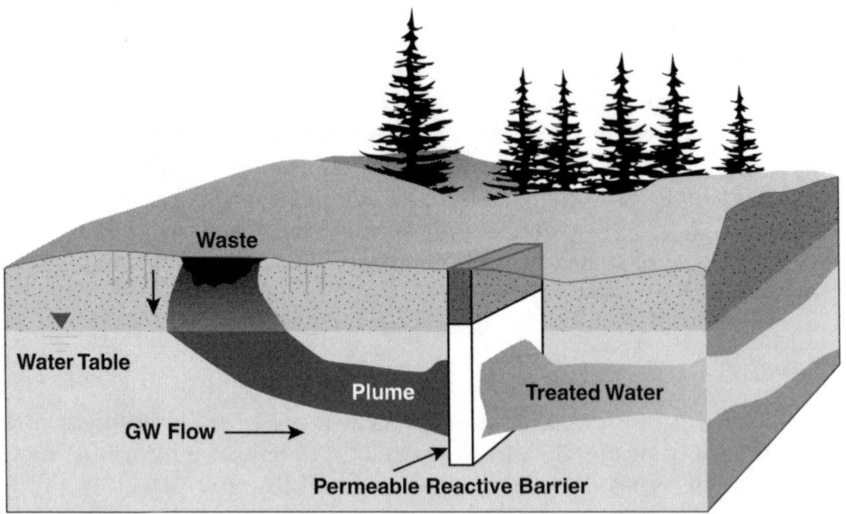

FIGURE 3-6 Schematic of a permeable reactive barrier. NOTE: GW = groundwater. SOURCE: EPA, 1998.

"immobilization" is actually only a retardation, and the conditions that were responsible for immobilization must be maintained. For example, if Cr and U are immobilized by reduction, and oxygen is subsequently reintroduced to the barrier, reoxidation would be slow for Cr but more rapid for U.

Application

Several approaches for installing the reactive treatment zone are possible. One approach, limited to shallow depths, is to excavate and backfill a trench with the reactive material, often in one pass. A second approach is to use slurry wall construction technology to create a larger and deeper permeable curtain. In this approach, a polymer mixed with reactive material replaces subsurface materials as excavation proceeds. When excavation is complete, the polymer is removed by pumping and biodegradation, leaving a permeable wall that contains the reactive material. A third approach is to install temporary sealable sheet piling to allow dewatering and installation of a reactive zone. A fourth approach is to inject the reactive material directly with a jet. For the first three approaches, costs are likely to be high if a continuous zone is installed across zones of contaminated water. A promising alternative is to use sealable piling to funnel the natural groundwater flow through narrow zones that contain the reactive material. This method allows greater control of the treatment zone and facilitates removal or replacement of the reactive material (NRC, 1994).

Performance

Table 3-10 provides examples of permeable reactive barrier installations with various reactive media. One of the most carefully studied reactive barriers is at Elizabeth City, North Carolina. A full-scale field demonstration of an Fe^0 reactive barrier to intercept a Cr(VI)

TABLE 3-10 Summary of Selected Permeable Reactive Barrier Installations for Treating Metals and Radionuclides

Reactive Medium	Contaminant	Study Type	Site	Reference
Fe^0	Cr	Commercial	Elizabeth City, N.C.	EPA, 1995b; RTDF, 1999; Puls et al., 1998
	Cr	Field	Elizabeth City, N.C.	EPA, 1995b; Sabatini et al., 1997; Puls et al., 1998
	U	Field	Durango, Colo.	Dwyer et al., 1996
	U	Field	Fry Canyon, Utah	RTDF, 1999; Naftz, 1997
Lime or limestone	Acid mine drainage	Commercial	Various sites	Kleinmann et al., 1983
	Pb, Cd, As, Zn, Cu	Commercial	Nesquehoning, Pa.	RTDF, 1999
$Fe(OH)_3$	U	Lab	Monticello, Utah	Morrison and Spangler, 1993; Morrison et al., 1995
	U	Pilot	Monticello, Utah	
	U	Field	Fry Canyon, Utah	RTDF, 1999; Naftz, 1997
Zeolites	Sr	Lab		Fuhrmann et al., 1995
Modified zeolites	Cr	Lab		Haggerty and Bowman, 1994; RTDF, 1999
Bentonite	Cs	Lab		Oscarson et al., 1994
Peat	Cr	Lab		Ho et al., 1995
	U	Lab		Morrison and Spangler, 1992
PO_4	U	Field	Fry Canyon, Utah	RTDF, 1999
Organic carbon	Ni, Fe	Commercial	Sudbury, Ontario	RTDF, 1999

SOURCE: Adapted from Vidic and Pohland, 1996; RTDF, 1999.

and trichloroethylene (TCE) plume is operating at a Coast Guard air station at this site. The mixed waste contaminant plume is between 4.3 and 6.1 m (14 and 20 ft) below ground surface, and the water table ranges from 1.5 to 1.8 m (5 to 6 ft) below ground surface. Chromium(VI) concentrations range as high as 28 mg/liter near the contaminant source. A pilot-scale demonstration began at this site in September 1994, and the full-scale field test began in June 1996. For the pilot test, 21 20-cm (8-in.) holes were installed in a staggered three-row array over a 5.6 m^2 (60 ft^2) area. A mixture of 50 percent iron filings, 25 percent clean coarse sand, and 25 percent aquifer material (by volume) was poured down the hollow stem augers to a depth of 3 to 6.7 m (10 to 22 ft) below ground surface (EPA, 1995b). The full-scale test involves a trench that was excavated and simultaneously backfilled with the reactive medium. The barrier is 0.6 m thick, 46 m long, and 7.3 m deep.

Chromium(VI) concentrations in the effluent from the full-scale barrier have been decreased to below detection (<0.01 mg/liter). Under the highly reducing conditions that prevail within the wall, the reduction of Cr(VI) to Cr(III) and the formation of an insoluble precipitate constitute the likely mechanism causing the contamination decrease.

Limitations

Because metals and radionuclides are nondegradable, treatment by sorption or precipitation within a reactive barrier must be regarded as a retardation of contaminant migration rather than as a permanent solution to the problem. If retardation is accomplished through reduction to an immobile form, either the reducing conditions must be maintained to prevent remobilization or the reduction reaction must be effectively irreversible. If sorption is responsible for the retardation, the degree of reversibility of the reaction must be considered.

Another limitation is that because of the difficulties of emplacing the barrier, barriers are generally limited to near-surface contamination. In addition, the long-term performance of reactive barriers remains an open question. Principal concerns are reduction of permeability of the barrier due to buildup of reaction products or passivation of the reactive surface of the iron.

Advantages

Permeable barriers require little or no energy input once installed and thus can result in lower overall treatment costs. Because the reaction zone is limited in area, it may be easier to design, monitor, maintain, and control than in systems operating over larger areas. Another strong advantage is the ability of this technology to treat contaminant mixtures (see Chapter 4).

In Situ Redox Manipulation

Description

In situ redox manipulation is the injection of chemical reductants into the ground or the stimulation of naturally occurring iron-reducing bacteria with nutrients in order to create reducing conditions in the subsurface, leading to reduction and immobilization of certain contaminants in groundwater (see Figure 3-7) (Amonette et al., 1994; Fruchter et al., 1997). This type of technology can be viewed as a special type of permeable reactive barrier, in which a part of the subsurface is transformed to a containment treatment zone.

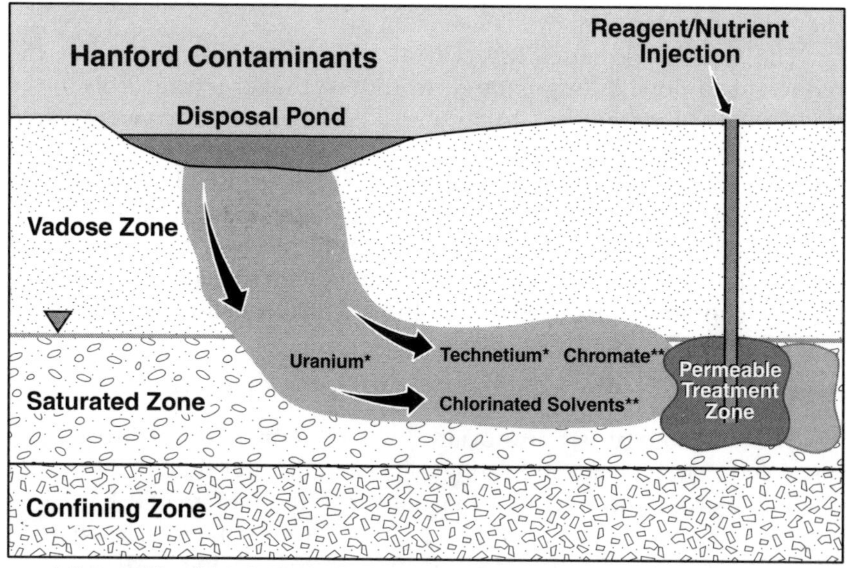

FIGURE 3-7 Schematic of an in situ redox manipulation system. SOURCE: Fruchter et al., 1997.

Application

This technology was conceived primarily for selected priority metal and radionuclide contaminants that are mobile in their oxidized form but immobile in the reduced form (Cr, U, Tc, Pu). Certain halogenated organic contaminants, including TCE and other chlorinated solvents, in theory can be treated at the same time (Betts, 1998).

Physical and Chemical Principles

The primary redox buffer in the reducing zone created by in situ redox manipulation systems is generally structural iron—that is, iron bound in clay interlayers of the aquifer material. This reduced iron forms a large reducing buffer region that is not reoxidized easily by oxygen, yet reacts readily enough with the contaminants to retard their transport. The elements Cr, Tc, U, and Pu can be reduced by mineral-bound Fe(II) to form very insoluble oxides. Various reductants have been tested, including N_2H_4, NH_2OH, SO_3^{2-}, S^{2-}, $S_2O_4^{2-}$, colloidal Fe(II) in clays, and Fe^0. In tests at the Hanford Site, dithionite ($S_2O_4^{2-}$) was most effective (Amonette et al., 1994). Lactate injection has been proposed as a method for stimulating iron-reducing bacteria (Fruchter et al., 1997), but field tests to date have focused on chemical reductants.

The reaction kinetics are critical. First, it is necessary for the reductant to persist long enough to reduce the structural iron in the clays but not long enough to become a contaminant of concern itself. Second, the immobile structural iron must be sufficiently reactive with the contaminants that they are in fact reduced and immobilized, but it must not be so reactive with other oxidants, such as dissolved oxygen, that the iron is re-oxidized before it reacts with the contaminants. Third, the reoxidation of the entire system must be slow enough that the contaminants are not remobilized after having been reduced.

Application

The technology is appropriate for contaminants that are widely dispersed in the unsaturated zone or in groundwater that is not readily accessible from the surface—that is, more than 15 m (50 ft) below ground surface. Liquid or gaseous chemical reductants, or substrate and nutrients to stimulate microbial growth, are injected to create reducing zones. The reductant must be easy to inject to the desired treatment depth via an injection well and must be acceptable to regulatory agencies.

Performance

Three field tests related to the use of in situ redox manipulation for Cr(VI) treatment have been conducted at the Hanford 100-H site: a bromide tracer test, a mini injection-withdrawal test, and a full-scale injection-withdrawal test (PNNL, 1996; Fruchter et al., 1997) (see Table 3-11). A large-scale treatability test is now under way at the Hanford 100-D site (Betts, 1998).

In the full-scale injection-withdrawal test, approximately 77,000 liters of 0.1 M sodium dithionite in a 0.44 M potassium bicarbonate-carbonate buffer at pH 11.2 were injected into an unconfined sandy-gravel aquifer of the Hanford formation. After the 18-hour injection period and 83-hour reaction period, approximately five injection volumes were withdrawn to recover unused reagent; 87 percent of the injected dithionite was recovered (Fruchter et al., 1997). The objectives were to create a reducing zone approximately 15 m in diameter and to monitor the removal of Cr(VI) and the lifetime of the reduced zone. The depth of the test was approximately 15 m (50 ft). The test was conducted with a single injection well and 15 monitoring wells. The target contaminant was chromate.

Monitoring data indicated that a year after the test, chromium levels in groundwater had decreased from an initial value of 60 μg/liter to below the detection limit (Fruchter et al., 1997). From 60 to 100 percent of the iron in the sediments was reduced by dithionite. No significant plugging of the aquifer formation (a potential problem

TABLE 3-11 Summary of In Situ Redox Manipulation Tests

Reactive Medium	Contaminants	Study Type	Site	Reference
Dithionite injection to reduce structural iron	Cr	Field	Hanford 100-H_f full scale	PNNL, 1996; Fruchter et al., 1997
		Field	Hanford 100-H_f push-pull[a]	Vermeul et al., 1995; Fruchter et al., 1996
		Intermediate	Physical model of 7-m-radius; 10-degree wedge of contaminated aquifer	Fruchter et al., 1996

[a] A push-pull test uses a single well for both injection of reactive agents and withdrawal of water samples

due to precipitation reactions) occurred. Data from after the test injection indicated that the aquifer remains reducing and chromate remains below the 8-μg/liter detection limit (Fruchter et al., 1998). The lifetime of the reducing zone has been estimated at 10 years.

Because of the success of this field test, a large-scale demonstration of in situ redox manipulation is now under way at Hanford. In this demonstration, a treatment zone approximately 50 m (150 ft) long is being created by overlapping cylindrical reduced zones created with five injection wells (Fruchter et al., 1998). Bench-scale tests are also under way at Hanford and elsewhere to test the performance of this method for treating chlorinated solvents.

Limitations

For application in relatively deep aquifers, verifying that the manipulated zone intercepts the contaminant plume will remain an uncertainty. Furthermore, the kinetics of all the reactions must be appropriate, as outlined above. Finally, this containment technology must be maintained far into the future to avoid reoxidation and mobilization of the contaminant. Reoxidization may not be a problem for Cr(VI) that has been reduced to Cr(III), because the reaction may be irreversible (or have very slow kinetics). However, reoxidation is a concern for other contaminants, such as U, for which the reaction is reversible.

Advantages

A prime advantage of in situ redox manipulation is the ability to treat contamination at depths that are inaccessible by excavation of any sort. The technology also is relatively inexpensive to install and operate. It allows management of large quantities of water in situ, avoiding the safety and regulatory issues associated with bringing water to the surface.

Bioremediation

Description

Bioremediation is usually associated with the microbiological degradation of organic contaminants to more benign forms. As applied to inorganic contaminants, bioremediation refers to processes through which contaminants are *mobilized* or *immobilized* as a direct result of microbiological activity. Mobilization can occur through complexation of an inorganic contaminant by soluble biologically produced complexing agents such cyclodextrans and exopolysaccharides or by reductive dissolution of metal oxides ("microbial leaching"); these mobilization processes are discussed later in this chapter. Immobilization, discussed here, can occur through reduction to an insoluble form, for example, Cr(VI) to Cr(III), U(VI) to U(IV), Pu(V, VI) to Pu(III, IV), and Tc(VII) to Tc(IV); immobilization may be enhanced by sorption to biomass (Ahmann, 1997).

Physical and Chemical Principles

Respiratory microorganisms obtain energy from the enzymatically mediated oxidation of a substrate (e.g., acetate, glucose, H_2) coupled to the reduction of a terminal electron acceptor (e.g., O_2, NO_3^-, Fe(III), SO_4^{2-}). Several toxic metals and radionuclides (including U, Pu, and Cr) have been shown to be reduced during this process, either directly as the terminal electron acceptor in the metabolic process or indirectly. If reducing conditions can be maintained by the addition of substrate and suitable nutrients, inorganic contaminants will remain in their highly insoluble, immobile forms.

Application

No definitive field studies have been reported specifically on manipulating the subsurface environment to cause microbiological reduction and immobilization of metals and radionuclides. Presumably, a field test could be designed similar to a field test for microbial treatment of chlorinated solvents by reductive dehalogenation, as described in Chapter 4. Examples of laboratory tests are given in Table 3-12.

Performance

Laboratory tests (see Table 3-10 and Tucker, 1996) have indicated that the immobilization of metals and radionuclides by bioremediation could be very effective, with removal of contaminants from the mobile aqueous phase to below critical values.

Limitations

Contaminant immobilization caused by bioremediation must be regarded as the retardation of contaminant migration rather than as a permanent solution to the problem. Reducing conditions that favor the immobilization reactions may have to be maintained to prevent remobilization.

TABLE 3-12 Selected Laboratory Tests of Microbiological Reduction and Immobilization of Metals and Radionuclides

Contaminant	Reference
Cr	Cifuentes et al., 1996 Turick et al., 1996 Losi et al., 1994a
U	Lovley et al., 1991 Lovley and Phillips, 1992 Sheppard and Evenden, 1992
Pu	Rusin et al., 1994 Zorpette, 1996 Macaskie, 1991

SOURCE: Adapted from Ahmann, 1997.

Advantages

Biological immobilization of metals and radionuclides could be performed simultaneously with bioremediation of organic compounds. The treatment occurs in situ, which decreases exposure risks and disposal problems. Costs are moderate.

TECHNOLOGIES FOR MOBILIZING AND EXTRACTING METALS AND RADIONUCLIDES

In addition to being treated by immobilization, some metals and radionuclides in groundwater and soil can be mobilized and extracted from the subsurface for treatment or disposal at the surface. Electrokinetic, soil flushing, soil washing, and phytoremediation processes, discussed below, are the primary technologies being developed for this purpose.

Electrokinetic Processes

Description

The application of an electric field to soil to remove chemical contaminants is called electrokinetic remediation. This process is particularly attractive for application to low-permeability soils that are difficult to flush. The reactions can be used to stabilize the contaminants in situ, or contaminants concentrated near the electrodes can be removed and treated ex situ.

In this process, a series of electrodes are placed into the contaminated area, and a 50- to 150-V direct current potential is applied between the electrodes (EPA, 1997b). The potential field causes movement of water and migration of the contaminants toward the electrode of opposite charge. Four processes are responsible for contaminant movement (Acar et al. 1995; EPA, 1997b):

1. electromigration (transport of charged chemical species in the electric gradient);
2. electroosmosis (transport of water or added pore fluid in the electric gradient);
3. electrophoresis (transport of charged particles in the electric gradient); and
4. electrolysis (chemical reactions at the electrodes resulting from the applied electrical potential).

Both electromigration to desorb and move anions and cations from the soil and transport them to the electrodes, and electroosmosis to drive a flushing fluid between the anode and cathode, have been used as the basis for electrokinetic remediation (EPA, 1997b). The contaminants removed with water or processing fluid can be treated ex situ by conventional processes.

Application

Electrokinetic processes have been used for the remediation of soils containing a number of inorganic contaminants (EPA 1997b,c). Removal of many contaminants, including cadmium, cesium, chromium, copper, lead, mercury, nickel, strontium, uranium, and zinc, has been demonstrated. Formation of complex thorium species may allow thorium removal.

An electrokinetic system at the Savannah River Site removed mercury and uranium. Ions were trapped in ion exchange polymer matrices in the electrode compartments (EPA, 1997b). In a bench-scale electrokinetic test conducted under the EPA SITE program, uranium, but not radium or thorium, was removed from kaolinite. Uranium precipitated as the hydroxide. The introduction of acetic acid into the cathode compartment prevented its precipitation near the cathode.

Limitations

In water, electrolysis produces acid at the anode and hydroxide at the cathode. The pH can drop to less than 2 at the anode and increase to more than 12 at the cathode. Unless movement of the acid front is retarded, the transport of hydrogen ion will predominate (Acar and Alshawabkeh, 1993). Transport of the acid front can be limited by the cation exchange capacity of the soil and by reaction with organic materials (e.g. humic acid) and inorganic compounds (e.g. calcium carbonate). Alternatively, water from one electrode can be extracted and reinjected at the other. The electrodes can also be placed in ceramic casings, which are kept filled with processing fluids chosen to maintain pH balance and assist in the solubilization and movement of contaminants (EPA, 1997b). The processing fluids can be pumped and the contaminants removed from them by precipitation or other treatment means.

Advantages

Electrokinetic systems can mobilize both metals and organic compounds as a result of the several processes that are responsible for contaminant movement.

Soil Flushing and Washing

Description

Soil flushing is an in situ process and soil washing is an ex situ process in which contaminants are removed from the soil by using a suitable extracting solution. Commonly used mobilizing agents are acids and chelating agents. Soil washing has been widely applied. Soil flushing has been used to recover metals in the mining industry, and considerable research has been conducted on the use of soil flushing for organic contaminants (see Chapter 4). However, this method has not been developed for the treatment of metals and radionuclides.

Physical and Chemical Principles and Application

Soil washing is generally applied after segregation of smaller-size (<63 µm) soil particles (NRC, 1997; Evanko and Dzombak, 1998). In general, the contaminant concentration is greater in smaller soil particles than in larger-size particles. Smaller particles preferentially bind contaminants as a consequence of their greater surface area and physicochemical reactivity.

A number of physical processes are available for pretreatment of the soil in soil washing systems (EPA, 1988; Smith et al., 1995; Evanko and Dzombak, 1998). Processes often employed include screen sizing, classification by settling velocity in air or water, gravity separation, flotation, and magnetic separation.

For soil washing, mobilizing chemicals are added to the separated soil in a reactor. For soil flushing, extractant chemicals are applied to the contaminated soil by surface flooding, sprinklers, leach fields, vertical or horizontal injection wells, basin infiltration systems, or trench infiltration systems (Evanko and Dzombak, 1998). After contact with the contaminated soil, the extractant is recovered for disposal or treatment and reuse.

Commonly used extractants include acids and chelating agents (Ehrenfeld and Bass, 1984; Rulkens and Assink, 1984; Smith et al., 1995). The most commonly used acids are sulfuric (H_2SO_4), hydrochloric (HCl), and nitric (HNO_3), and the most commonly used chelating agents are EDTA, citric acid, and DTPA (Smith et al., 1995). Although highly effective in metal mobilization, chelating agents such as EDTA are expensive and difficult to recover; however, Allen and Chen (1993) have reported that both EDTA and the contaminant metal can be recovered electrochemically. Because many metals are redox sensitive, oxidants and reductants have also been employed in mobilizing solutions. Soil flushing using water alone is often effective in removing hexavalent chromium because of its high solubility and mobility.

Ion exchange can be used to concentrate contaminants after recovery of mobilizing solutions. Other treatments for recovered solutions include evaporation and solidification, precipitation, coprecipitation, and sorption onto clay (IAEA, 1997). DOE has studied a number of conventional and advanced methods for the recovery of metals and radionuclides from water. Among these are ion exchangers attached to magnetic particles, semipermeable membranes, selective solid-phase extraction, and concentration with chelators.

Performance

A number of commercial vendors offer soil washing technology (EPA, 1995a). The technology has been demonstrated numerous times as an effective ex situ method for treating soil contaminated with both metals and organic contaminants. For example, in a SITE demonstration of a soil washing and chemical treatment process at the Twin Cities Army Ammunition Plant Site F in Minnesota, the process reduced lead concentrations from initial levels of 3,000 to 10,000 ppm (parts per million) to treated levels of less than 300 ppm.

In a recent field test of an in situ soil flushing system in the Province of Utrecht, The Netherlands, remediation was conducted by infiltrating acidified water (0.001 M HCl) into the subsurface (Otten et al., 1997). The treated area was 6,000 m^2 and 4-5 m deep and was contaminated with an estimated 725 kg of cadmium at concentrations ranging from 5 to 20 mg/kg. Treatment reduced the Cd concentration to less than 2.5 mg/kg in most of the treated area, except in a small zone where the initial Cd concentration was too high.

Limitations

Soils with high concentrations of clay and silt may be difficult to treat because of the difficulty in removing more tightly bound contaminants. Mineralized metals and metal particles (i.e., nonionic) are not easily treated. Suitable washing solutions may be difficult to find for complex mixtures of contaminants. Soil flushing may mobilize chemicals that are difficult to recover. Also, soil flushing can be applied only in geologic formations with sufficient permeability to allow circulation and recovery of the flushing solution.

Advantages

Transfer of the contaminant to a liquid stream often facilitates its treatment. There are a large number of methods for the treatment of liquid waste streams, for example, those employed in industrial waste treatment. Soil flushing has the added advantage of requiring no excavation.

Phytoremediation

Description

Phytoremediation refers to the use of plants to extract metals and metalloids from contaminated soils as shown in Figure 3-8. In phytoremediation, contaminated soil is seeded with special plants known as "hyperaccumulators" that can take up large quantities of metals or radionuclides through their root systems. The plants are then grown and harvested. Multiple harvests generally are required to clean up the site.

Phytoremediation is the direct and/or indirect use of green plants for remediation of contaminated groundwater or soil. The method has been of interest for several years because of the many observed mechanisms by which plants can remove, degrade, or immobilize a wide variety of contaminants. Phytoremediation has been applied to organic compounds, as well as metals and radionuclides.

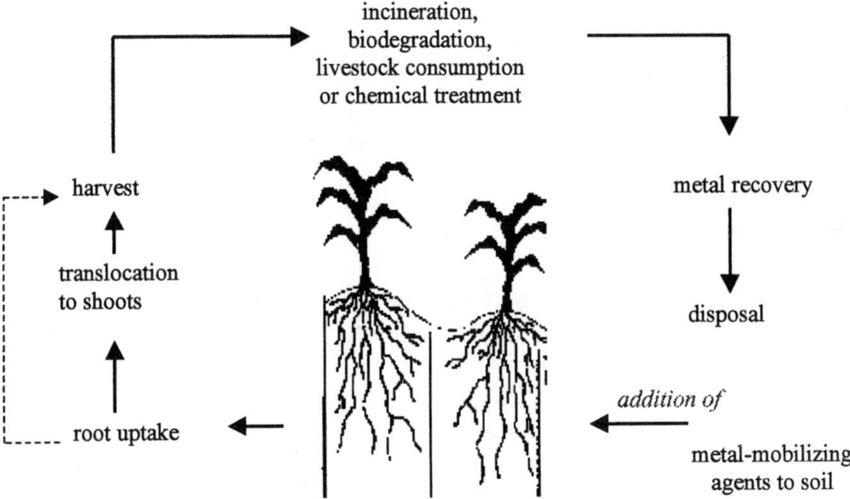

FIGURE 3-8 Phytoremediation of metals.

Phytoremediation can involve uptake of contaminants within the root zone and transport or accumulation within the plant. It also can involve enhancement of microbial processes within the soil for contaminant immobilization or degradation. In some cases it may involve degradation directly within the plant. The various types of processes have been described by the following somewhat overlapping terms (Schnoor, 1997; Dupont et al., 1998;):

- *Phytoextraction* involves the uptake of metals and radionuclides by plant roots, and their accumulation in the above-ground parts of the plant. The plant is harvested and processed to concentrate the contaminants.
- *Phytostabilization* refers to the use of vegetation to prevent the erosion of contaminanted soil, immobilize contaminants in the soil, or control groundwater movement through transpiration.
- *Rhizodegradation* involves stimulating microorganisms around the root zone, resulting in enhanced microbial degradation of the contaminants.
- *Phytodegradation* refers to the transformation of organic contaminants to less toxic compounds through their adsorption, uptake, or degradation by either the plant itself or plant-associated microflora.
- *Phytovolatilization* is a process by which contaminants can be taken up by the plant and then volatilized.

In general, phytoextraction and phytostabilization are the most important phytoremediation methods for treating contaminants in groundwater and soil (although phytovolatilization could be used to remove Hg). Phytostabilization is a special form of containment technology, whereas phytoextraction can be used to remove metals and radionuclides from the subsurface; thus, this discussion focuses on phytoextraction.

Physical and Chemical Principles

Many plants exude chelating agents from the roots to complex essential micronutrients and make them available to the plant. Phytoextraction causes plants to take up contaminant metals in the same way they extract micronutrients. For this method to be effective, some mechanism must exist for the plant to select a contaminant over other similar but more abundant metals (e.g., selection of Ra or Sr rather than Ca).

Application

Phytoextraction is most applicable to large areas of surface soils with low to moderate levels of contamination. Plants that accumulate at least 0.1 percent by weight of Co, Cu, Cr, Pb, or Ni or 1 percent by weight of Mn or Zn are defined as hyperaccumulators and are suitable for phytoextraction. Although many plants have been tested, plants of the genera *Brassica, Thlaspi, Cardaminopsis,* and *Alyssum* appear to be the most promising (Ahmann, 1997, after Kumar et al., 1995). For metals that are bound extremely tightly to soils, the addition of chelating agents, such as EDTA promotes their accumulation in the plant. In such cases, care must be exercised that adding the chelating agent does not cause leaching of the contaminant from the surface soil zone. Ultimately the plants are harvested, dried, and combusted or composted to reduce their mass prior to disposal. Multiple harvests are generally required to achieve cleanup goals.

Performance

Commercialization of phytoextraction and other phytoremediation methods has been relatively slow but appears to be occurring. Table 3-13 lists several field demonstrations of phytoextraction for the treatment of metals and radionuclides.

TABLE 3-13 Field Demonstrations of Phytoextraction of Metals and Radionuclides

Location	Application	Plants	Contaminants	Performance
Trenton, New Jersey	Demonstration on 60 x 90 m plot at brownfield site	*Brassica juncea* (Indian mustard)	Pb	Pb was removed to below action level in one season.
Pennsylvania	Mine waste site	*Thlaspi caerulescens*	Zn, Cd	Rapid contaminant uptake was observed, but the soil was difficult to decontaminate.
Findlay, Ohio	Contaminated soil demonstration under the SITE program	NA	Pb, Cr, Ni, Zn, Cd	NA

NOTE: NA = published information not available from the indicated sources.

SOURCES: Schnoor, 1997; Chappell, 1998.

Limitations

Phytoextraction is applicable only to the rooting zone of soils Although the goal of phytoextraction is mobilization of contaminants from soils to plant biomass for subsequent recovery or disposal, care must be exercised to prevent mobilization into the biosphere (for example, from animals eating the plants) where recovery is no longer possible. Furthermore, care must be exercised that natural or synthetic chelating agents associated with the uptake of contaminants do not mobilize the contaminant into groundwater. The plants must be able to grow vigorously at the site. Disposal of plant biomass can be a problem when plants are contaminated with heavy metals and radionuclides.

Advantages

Phytoextraction is a low-cost method to remove contaminants from large areas of the surface zone of contaminated soils.

CONCLUSIONS

Treatment technologies for cleaning up metal and radionuclide contaminants in groundwater and soil act either by immobilizing contaminants in place to prevent transport to humans and sensitive ecosystems or by mobilizing contaminants for extraction and treatment at the surface. Because metals and radionuclides, unlike organic contaminants, are nondegradable except by radioactive decay and because the risk posed by these compounds is highly sensitive to geochemical conditions, managing these contaminants is very different from managing the types of organic contaminants discussed in Chapter 4. Few well-established technologies are available for treating metals and radionuclides in the subsurface, but a number of new technologies are being developed. Available technologies for treating metals and radionuclides are summarized in Tables 3-4 and 3-5 and include the following:

- **Impermeable barriers** are among the least expensive and most widely used methods for preventing the spread of metal and radionuclide contaminants in groundwater. Vertical barriers are well developed and widely available; methods are being developed for the installation of horizontal barriers beneath existing waste.
- **In situ vitrification** for immobilization of metal and radionuclide contaminants is an emerging technology that is particularly suitable for sites with high concentrations of long-lived radioisotopes within 6 to 9 m

of the soil surface (depending on water table depth and soil moisture). However, it is among the most expensive treatment technologies.

- **Solidification and stabilization** are mature technologies for use ex situ but are considered emerging technologies for use in situ. Ensuring sufficient mixing is difficult when this technology is used in situ. Improved mixing methods are being tested. The longevity of the solidified or stabilized material is another concern that must be addressed before these methods can be considered established technologies.
- **Permeable reactive barriers** are among the most promising and rapidly developing emerging treatment technologies for metal and radionuclide contaminants. A variety of reactive media has been tested for a variety of contaminants, including organics. Because the technology is relatively new, the longevity of the barrier is a major uncertainty.
- **In situ redox manipulation** is an emerging method that is appropriate at both shallow depths and depths at which trenches are impractical. It is an excellent technology for elements (e.g., Cr) that can be reduced to solids that are resistant to reoxidation by ambient oxygen, but it is less suitable for elements (e.g., Tc) that are susceptible to reoxidation.
- **Bioremediation** is in the early stages of development for the treatment of metals and radionuclides. If better developed, it could be a relatively low-cost alternative and could be used to treat mixtures of organic and inorganic contaminants.
- **Electrokinetics** may be advantageous for extracting metals and radionuclides from media with very low hydraulic conductivity. The consequences of generating large amounts of acid and base must be considered. Extensive field tests of electrokinetics for the remediation of metal and radionuclide contamination have yet to be conducted in the United States.
- **Soil washing** has been well developed for ex situ treatment of contaminated, coarse-grained, near-surface soils but requires excavation of the soil prior to treatment.
- **Soil flushing** can potentially flush metals and radionuclides from soil in situ. Although used in the mining industry, this technology has not seen widespread application in the remediation of metals and radionuclides.
- **Phytoremediation** is primarily advantageous for extracting metals from large areas of contaminated surface soils. Its advantages are low cost and ease of implementation. Potential disadvantages are the excessive mobilization of metals from chelators that often must be added and the need to dispose of plant biomass.

In general, only containment and ex situ technologies are well developed for treating metal and radionuclide contaminants. Additional development work is needed to increase the range of options for treating metals and radionuclides in situ and for extracting them for ex situ treatment.

REFERENCES

Acar, Y. B., and A. N. Alshawabkeh. 1993. Principles of electrokinetic remediation. Environmental Science and Technology 27:2638-2647.

Acar, Y. B., R. J. Gale, A. N. Alshawabkeh, R. E. Marks, S. Puppala, M. Bricka, and R. Parker. 1995. Electrokinetic remediation: Basics and technology status. Journal of Hazardous Materials 40:117-137.

Adriano, D. C. 1986. Trace Elements in the Terrestrial Environment. New York: Springer-Verlag

Ahmann, D. 1997. Bioremedation of metal-contaminated soil. Society for Industrial Microbiology News 47:218-233.

Allen, H. E., and P-H. Chen. 1993. Remediation of metal contaminated soil by EDTA incorporating electrochemical recovery of metal and EDTA. Environmental Progress 12:284-293.

Allen, H. E., and Y. Yin. 1998. Combining chemistry and biology to derive soil quality criteria for pollutants. Presented at Sixth World Congress of Soil Science, Montpellier, France, August 25.

Allison, J. D., D. S. Brown, and K. J. Novo-Gradac. 1991. MINTEQA2/PRODEFA2, A geochemical assessment model for environmental systems: Version 3.0 users manual. EPA/600/3-91/021. Washington, D.C.: U.S. Environmental Protection Agency.

Amonette, J. E., J. Szecsody, J. C. Templeton, Y. A. Gorby, and J. S. Fruchter. 1994. Abiotic reduction of aquifer materials by dithionite: A promising in situ remediation technology. In In-Situ Remediation: Scientific Basis for Current and Future Technologies, G. W. Gee and N. R. Wing, Eds. Richland, Wash.: Battelle Press.

Ballard, J. H., and M. J. Cullinane, Jr. 1997. Tri-Service Site Characterization and Analysis Penetrometer System (SCAPS) Technology Verification and Transition. Vicksburg, Miss.: U.S. Army Corps of Engineers Waterways Experiment Station.

Betts, K. S. 1998. Novel barrier remediates chlorinated solvents. Environmental Science and Technology 3(2):495A.

Borns, D. J. 1997. Geomembranes with incorporated optical fiber sensors for geotechnical and environmental applications. Pp. 1067-1073 in Proceedings of the International Containment Technology Conference, St. Petersburg, Fla., February 9-12.

Browne, C. L., Y. M. Wong, and D. R. Buhler. 1984. A predictive model for the accumulation of cadmium by container-grown plants. Journal of Environmnetal Quality 13:184.

Buergisser, C. S., and A. T. Stone. 1997. Determination of EDTA, NTA, and other amino carboxylic acids and their Co(II) and Co(III) complexes by capillary electrophoresis. Environmental Science and Technology. 31:2656-2664.

Bufflap, S. E., and H. E. Allen. 1995. Sediment pore water collection methods: A review. Water Research 29:165-177.

Chappell, J. 1998. Phytoremediation of TCE in ground water using populus. Washington, D.C.: EPA.

Cifuentes, F. R., W. C. Lindemann, and L. L. Barton. 1996. Chromium sorption and reduction in soil with implications to bioremediation. Soil Science 161:233-241.

Daily, W., and A. Ramirez. 1997. A new geophysical method for monitoring emplacement of subsurface barriers. Pp. 1053-1059 in Proceedings of the International Containment Technology Conference, St. Petersburg, Fla., February 9-12.

Dash, J. G., H. Y. Fu, and R. Leger. 1997. Frozen soil barriers for hazardous waste confinement. Pp. 607-613 in Proceedings of the International Containment Technology Conference, St. Petersburg, Fla., February 9-12.

Davis, J. L., and A. P. Annan. 1989. Ground penetrating radar for high resolution mapping of soil and rock stratigraphy. Geophysical Prospecting, 37:531-551.

Dupont, R. R., C. J. Bruel, D. C. Downey, S. C. Huling, M. C. Marley, R. D. Norris, and B. Pivetz. 1998. Innovative site remediation technology: Design and application: Bioremediation. Annapolis, Md.: American Academy of Environmnetal Engineers.

Dwyer, B. D., D. C. Marozas, K. Cantrell, and W. Stewart. 1996. Laboratory and field demonstration of reactive barrier systems. In Proceedings of the 1996 Spectrum Conference, Seattle, Wash., August 18-23.

Dzombak, D. A., and F. M. M. Morel. 1990. Surface Complexation Modeling: Hydrous Ferric Oxide. New York: Wiley.

Ehrenfeld, J., and J. Bass. 1984. Evaluation of Remedial Action Unit Operations at Hazardous Waste Disposal Sites. Park Ridge, N.J.: Noyes Publications.

EPA (Environmental Protection Agency). 1988. Technological Approaches to the Cleanup of Radiologically Contaminated Superfund Sites. EPA/540/2-88/002. Washington, D.C.

EPA. 1992. Vitrification Technologies for Treatment of Hazardous and Radioactive Waste. EPA/625/R-92/002. Washington, D.C.: EPA, Office of Research and Development.

EPA. 1994. Innovative Site Remediation Technology: Solidification/Stabilization, Vol. 4 EPA 542/B-94/001. Washington, D.C.

EPA. 1995a. Contaminants and Remedial Options at Selected Metal-Contaminated Sites. EPA 540/R-95/512. Washington, D.C.

EPA. 1995b. In Situ Remediation Technology Status Report: Treatment Walls. EPA 542-K-94-004. Washington, D.C.: EPA, Office of Solid Waste and Emergency Response.

EPA. 1997a. Cleaning Up the Nation's Waste Sites: Markets and Technology Trends. 1996 Edition. EPA 542-R-96-005. Washington, D.C.: EPA, Office of Solid Waste and Emergency Response.

EPA. 1997b. Electrokinetic Laboratory and Field Processes Applicable to Radioactive and Hazardous Mixed Waste in Soil and Groundwater. EPA 402/R-97/006. Washington, D.C.

EPA. 1997c. Recent Developments for In Situ Treatment of Metal-Contaminated Soils. EPA 542/R-97/004. Washington, D.C.

EPA. 1997d. Best Management Practices (BMPS) for Soil Treatment Technologies: Suggested Guidelines to Prevent Cross-Media Transfer of Contaminants During Clean-Up Activities. EPA 530/R-97/007. Washington, D.C.

EPA. 1997e. Engineering Bulletin. Technology Alternatives for the Remediation of Soils Contaminated with Arsenic, Cadmium, Chromium, Mercury, and Lead. EPA 540/S-97/500. Washington, D.C.

EPA. 1998. Permeable Reactive Barrier Technologies for Containment Remediation. EPA/600/R-98/125. Washington, D.C.: EPA.

Evanko, C. R., and D. A. Dzombak. 1998. Remediation of Metals-Contaminated Soils and Ground-Water. Technology Evaluation Report. Pittsburgh, Pa.: Ground-Water Re–mediation Technologies Analysis Center.

Evans, J. C., M. Allan, S. R. Day, G. M. Filz, H. I. Inyang, S. Jefferis, L. E. Kukacka, L. Martinenghi, R. Mitchell, and K. Potter 1996. Soil- and cement-based vertical barriers with focus on materials. In Assessment of Barrier Containment Technologies, R. E. Rumer and J. K. Mitchell, Eds. PB96-180583. Spingfield, Va.: National Technical Information Service.

Filz, G. M., J. K. Mitchell., L. R. Anderson, J. D. Betsill, E. E. Carter, R. R. Davidson, S. R. Day, A. Esnault, J. C. Evans, S. Jefferis, M. Manassero, L. Martinenghi, R. L. Stammes, G. J. Tamaro, G. A. M. van Meurs, and D. S. Yang. 1996. Design, construction and performance of soil- and cement-based vertical barriers. In Assessment of Barrier Containment Technologies, R. E. Rumer and J. K. Mitchell, Eds. #PB96-180583. Springfield, Va.: National Technical Information Service.

Fish, W., D. A. Dzombak and F. M. M. Morel. 1986. Metal-humate interactions. 2. Application and comparison of models. Environmental Science and Technology 20: 676-683.

Florence, T. M. 1989. Electrochemical techniques for trace element speciation in waters. In Trace Element Speciation: Analytical Methods and Problems, G. E. Batley, Ed. Boca Raton, Fla.: CRC Press.

Fruchter, J. S., J. E. Amonette, C. R. Cole, Y. A. Gorby, M. D. Humphrey, J. D. Istok, F. A. Spane, J. E. Szecsody, S. S. Teel, V. R. Vermeul, M. D. Williams, and S. B. Yabusaki. 1996. In Situ Redox Manipulation Field Injection Test Report-Hanford 100H Area. PNNL-11372. Richland, Wash.: Pacific Northwest National Laboratory.

Fruchter, J. S., C. R. Cole, M. D. Williams, V. R. Vermeul, S. S. Teel, J. E. Amonette, J. E. Szecsody, and S. B. Yabusaki. 1997. Creation of a subsurface permeable treatment barrier using in situ redox manipulation. Pp. 704-710 in Proceedings, International Containment Technology Conference, February 9-12. Washington, D.C.: Department of Energy.

Fruchter, J. S., C. R. Cole, M. D. Williams, V. R. Vermeul, J. E. Szecsody, and J. C. Evans. 1998. Recent progress on in situ redox manipulation barriers for chromate and trichloroethylene. Presented at Fifth Meeting of the NRC Committee on Technologies for Cleanup of Subsurface Contamination in the DOE Weapons Complex, Richland, Wash., May 14-15.

Fuhrmann, M., D. Aloysius, and H. Zhou. 1995. Permeable, subsurface sorbent barrier for ^{90}Sr: Laboratory studies of natural and synthetic materials. In Proceedings of Waste Management '95, Tucson, Az., February 16-March 2.

Haggerty, G. M., and R. S. Bowman. 1994. Sorption of chromate and other inorganic anions by organo-zeolite. Environmental Science and Technology 28:452-458.

Hajós, P., G. Révész, and O. Horváth. 1996. The simultaneous analysis of metal-EDTA complexes and inorganic anions by suppressed ion chromatography. Journal of Chromatographic Science 34:291-299.

Ho, Y. S., D. A. Wase, and C. G. Forster. 1995. Batch nickel removal from aqueous solution by sphagnum moss peat. Water Research 29:1327-1332.

IAEA (International Atomic Energy Agency). 1997. Technical Options for the Remediation of Radioactively Contaminated Groundwater. Vienna: IAEA.

Inyang, H. I., J. D. Betsill, R. Breeden, G. H. Chamberlain, S. Dutta, L. Everett, R. Fuentes, J. Hendrickson, J. Koutsandreas, D. Lesmes, G. Loomis, S. M. Mangion, C. Pfeifer, R. W. Puls, R. L. Stamnes, T. D. Vandel, and C. Williams. 1996. Performance monitoring and evaluation. In Assessment of Barrier Containment Technologies, R. E. Rumer and J. K. Mitchell, Eds. PB96-180583. Springfield, Va.: National Technical Information Service.

Josten, N. E., R. J. Gehrke, and M. V. Carpenter. 1995. Dig-Face Monitoring During Excavation of a Radioactive Plume at Mound Laboratory, Ohio. INEL-9510633. Idaho Falls: Idaho National Engineering and Environmental Laboratory.

Kheboian, C., and C. F. Bauer. 1987. Accuracy of selective extraction procedures for metal speciation in model aquatic sediments. Analytical Chemistry 59:1417-1423.

Kleinmann, R. L. P., T. O. Tiernan, J. G. Solch, and R. L. Harris. 1983. A low-cost, low-maintenance treatment system for acid mine drainage using sphagnum moss and limestone. In National Symposium on Surface Mining, Hydrology, Sedimentology and Reclamation. Lexington, Kentucky: University of Kentucky.

Koerner, R. M., J. L. Guglielmetti, R. C. Bachus, P. T. Burnette, N. Cortlever, J. M. Cramer, R. E. Landreth, S. M. Mangion, M. Phifer, and W. M. Walling. 1996. Vertical barriers: Geomembranes. In Assessment of Barrier Containment Technologies, R. E. Rumer and J. K. Mitchell, Eds. PB96-180583. Springfield, Va.: National Technical Information Service.

Korte, N., O. R. West, F. G. Gardner, S. R. Cline, J. Strong-Gunderson, R. L. Giegrist, and J. Baker. 1997. Deep soil mixing for reagent delivery and contaminant treatment. Pp. 525-530 in Proceedings of the International Containment Technology Conference, St. Petersburg, Fla., February 9-12.

Kumar, P. B., V. Dushenkov, H. Motto., and I. Raskin. 1995. Phytoextraction: The use of plants to remove heavy metals from soils. Environmental Science and Technology 29:1232-1238.

Lesmes, D., D. Cist, and D. Morgan. 1997. Ground penetrating radar investigation of a frozen earth barrier. Pp. 1074-1080 in Proceedings of the International Containment Technology Conference, St. Petersburg, Fla., February 9-12.

Losi, M. E., C. Amrhein, and W. T. Frankenberger, Jr. 1994. Factors affecting chemical and biological reduction of hexavalent chromium in soil. Environmental Toxicology and Chemistry 13:1727-1735.

Lovley, D. R., and E. J. P. Phillips. 1992. Bioremediation of uranium contamination with enzymatic uranium reduction. Environmental Science and Technology 26:2228-2234.

Lovley, D. R., E. J. Phillips, Y. A. Gorby, and E. R. Landa. 1991. Microbial reduction of uranium. Nature 350:413-416.

Macaskie, L. E. 1991. The application of biotechnology to the treatment of wastes produced from the nuclear fuel cycle: Biodegradation and bioaccumulation as a means of treating radionuclide containing streams. Critical Reviews in Biotechnology 11:41-112.

Martin, J. P., P. Nire, and A. J. Thomas. 1987. Sequential extraction techniques: Promises and problems. Marine Chemistry 22:313-342.

McMahon, D. R., R. Fuentes, E. W. Gleason, B. M. LaRue, P. C. Repetto, D. J. A. Smyth, and A. Street. 1996. Vertical barriers: Sheet piles. In Assessment of Barrier Containment Technologies, R. E. Rumer and J. K. Mitchell, Eds. PB96-180583. Springfield, Va.: National Technical Information Service.

Morrison, S. J., and R. R. Spangler. 1992. Extraction of uranium and molybdenum from aqueous solution: A survey of industrial materials for use in chemical barriers for uranium mill tailings remediation. Environmental Science and Technology 26:1922-1931.

Morrison, S. J., and R. R. Spangler. 1993. Chemical barriers for controlling ground water contamination. Environmenatl Progress 12:175-181.

Morrison, S. J., V. S. Tripathi, and R. R. Spangler. 1995. Coupled reaction transport modeling of a chemical barrier for controlling uranium(VI) contamination in groundwater. Journal of Contaminant Hydrology 17:347-363.

Naftz, D. L. 1997. Field demonstration of reactive chemical barriers to control radionuclide and trace-element contamination in ground water, Fry Canyon, Utah. Presented at 1997 GSA Annual Meeting, Salt Lake City, October 20-23.

Nordstrom, D. K., L. N. Plummer, T. M. L. Wigley, T. J. Wolery, J. W. Ball, E. A. Jenne, R. L. Bassett, D. A. Crerar, T. M. Florence, B. Fritz, M. Hoffman, G. R. Holdren, Jr., G. M. Lafon, S. V. Mattigod, R. E. McDuff, F. Morel, M. M. Reddy, G. Sposito, and J. Thraikill. 1979. Comparison of computerized chemical models for equilibrium calculations in aqueous systems. In Chemical Modeling in Aqueous Systems: Speciation, Sorption, Solubility and Kinetics, E. A. Jenne, Ed. Symp. Ser. 93. Washington, D.C.: American Chemical Society.

NRC (National Research Council). 1994. Alternatives for Ground Water Cleanup. Washington, D.C.: National Academy Press.

NRC. 1997. Innovations in Ground Water and Soil Cleanup. Washington, D.C.: National Academy Press.

Oscarson, D. W., H. B. Hume, and F. King. 1994. Sorption of cesium on compacted bentonite. Clays and Clay Minerals 42:731-736.

Otten, A., A. Alphenaar, C. Pijls, F. Spuij, and H. de Wit. 1997. In Situ Soil Remediation. Boston: Kluwer Academic Publishers.

Pellerin, L. 1997. Geophysical verification of the thin diaphragm wall barriers at the Dover National Test Site. Final report submitted by LBNL to the DOE-EM SCFA. Berkeley, Calif.: Lawrence Berkeley National Laboratory.

Perdue, E. M., and C. R. Lytle. 1983. Distribution model for binding of protons and metal ions by humic substances. Environmental Science and Technology 17:654-660.

Persoff, P, G., J. Moridis, J. Apps, K. Pruess, and S. J. Muller. 1994. Designing injectable colloidal silica barriers for waste isolation at the Hanford Site. Pp. 87-102 in In-Situ Remediation: Scientific Basis for Current and Future Technologies, G. W. Gee and N. R. Wing, Eds. Richland, Wash.: Battelle Press.

Peters, R. 1994. Demonstration of ground freezing for radioactive/hazardous waste disposal. Pp. 103-112 in In-Situ Remediation: Scientific Basis for Current and Future Technologies, G.W. Gee and N.R. Wing, Eds. Richland, Wash.: Battelle Press.

Peterson, M. E., R. C. Landis, G. Burke, M. Cherrington, B. Dwyer, B. Gemmi, A. Iskandar, G. Loomis, R. Peters, R. Waters, and P. Yen. 1996. Artificially Emplaced Floors and Bottom Barriers. In Assessment of Barrier Containment Technologies, R. E. Rumer and J. K. Mitchell, Eds. PB96-180583.

PNNL (Pacific Northwest National Laboratory). 1996. In Situ Redox Manipulation Field Injection Test Report—Hanford 100 H Area. PNNL-11372. Richland, Wash.: Battelle Pacific Northwest National Laboratory.

Puls, R. W., D. W. Blowes, and R. W. Gillham. 1998. Emplacement verification and long-term performance monitoring for permeable reactive barrier at the USCG support center, Elizabeth City, North Carolina. Pp. 459-466 in Groundwater Quality: Remediation and Proctection. PB98-151285. Springfield, Va.: National Technical Information Service.

Rabideau, A. J., C. B. Andrews, C. Chiang, P. Grathwol, J. S. Hayworth, P. Culligan-Hensley, M. Manassero, J. W. Mercer, H. Mott, P. R. Schroeder, C. Shackelford, G. Teutsch, G. A. M. van Meurs, R. Wilhelm, and C. Zheng. 1996. Containment transport modeling. In Assessment of Barrier Containment Technologies, R. E. Rumer and J. K. Mitchell, Eds. PB96-180583. Springfield, Va.: National Technical Information Service.

Radovanovic, H., and A. A. Koelmans. 1998. Prediction of in situ trace metal distribution coefficients for suspended solids in natural waters. Environmental Science and Technology 32:753-759.

Rapin, F., A. Tessier, P. G. C. Campbell, and R. Carignan. 1986. Potential artifacts in the determination of metal partitioning in sediments by a sequential extraction procedure. Environmental Science and Technology 20:836-840.

Rendall, P. S., G. E. Batley, and A. J. Cameron. 1980. Adsorption as a control of metal concentrations in sediment extracts. Environmental Science and Technology 14:314-318.

Riley, R. G., J. M. Zachara, and F. J. Wobber. 1992. Chemical Contaminants on DOE Lands and Selection of Contaminating Mixtures for Subsurface Science Research. DOE/ER-0547T. Washington, D.C.: DOE, Office of Energy Research.

RTDF (Remediation Technologies Development Forum). 1999. Permeable Reactive Barrier Installation Profiles. Washington, D.C.: EPA. http://www.rtdf.org/public/permbarr/barrdocs.htm.

Rulkens, W. H., and J. W. Assink. 1984. Extraction as a method for cleaning contaminated soil: Possibilities, problems and research. Pp. 576-583 in Proceedings of the 5th National Conference on Management of Uncontrolled Hazardous Waste Sites, Washington, D.C.

Rumer, R. R., and J. K. Mitchell. 1995. Assessment of Barrier-Containment Technologies: A Comprehensive Treatment for Environmnetal Remeditaion Applications. Springfield, Va.: National Technical Information Service.

Rumer, R. R., and M. E. Ryan. 1995. Barrier Containment Technologies for Environmental Remediation Applications. New York: Wiley.

Rusin, P., L. Quintana, J. Brainard, B. Streitlemeier, C. Tait, S. Ekberg, P. Palmer, T. Newton, and D. Clark. 1994. Solubilization of plutonium hydrous oxide by iron reducing bacteria. Environmental Science and Technology 28:1686-1690.

Russel, K., and A. Rabideau. 1997. Impact of vertical barriers on performance of pump-and-treat systems. Pp. 902-909 in Proceedings of the International Containment Technology Conference, St. Petersburg, Fla., February 9-12..

Sabatini, D. A., R. C. Knox, E. E. Tucker, and R. W. Puls. 1997. Environmental Research Brief: Innovative Measures for Subsurface Chromium Remediation: Source Zone, Concentrated Plume, and Dilute Plume. EPA/600/5-971005. Washington, D.C.: EPA.

Schecher, W. D., and D. C. McAvoy. 1992. MINEQL+: A software environment for chemical equilibrium modeling. Computers, Environment and Urban Systems 16:65.

Schnoor, J. L. 1997. Phytoremediation. Pittsburgh, Pa.: Ground-Water Remediation Technologies Analysis Center.

Schultz, D. S., and R. C. Landis. 1998. Design and cost estimation of permeable reactive barriers. Remediation 9(1):57-68.

Sheppard, M. I., and M. Stephenson. 1997. Critical evaluation of selective extraction methods for soils and sediments. Pp. 69-97 in Contaminated Soils: Proceedings of 3rd International Conference on the Biogeochemistry of Trace Elements. R. Prost, Ed. Paris: INRA.

Sheppard, M. I. and D. H. Thibault. 1992. Desorption and extraction of selected heavy metals from soils. Journal of the Science Society of America 56:514-523.

Sheppard, S. C., and W. G. Evenden. 1992. Bioavailability indices for uranium: Effect of concentration in eleven soils. Archives of Environmental Contamination and Toxicology 23:117-124.

Siegrist, R. L., O. R. West, M. I. Morris, D. A. Pickering, D. W. Greene, C. A. Muhr, D. D. Davenport, and J. S. Gierke. 1995. In situ mixed region vapor stripping in low-permeability media. 2. Full-scale field experiments. Environmental Science and Technology 29(9):2198-2207.

Smith, L. A., J. L. Mearis, A. Chen, B. Alleman, C. C. Chapman, J. S. Tixier, Jr., E. Brauning, A. R. Gavaskar, and M. D. Royet. 1995. Remedial Options for Metals-Contaminated Sites. Boca Raton, Fla.: Lewis Publishers.

Smyth, D. J. A., S. G. Shikaze, and J. A. Cherry. 1997a. Hydraulic performance of permeable barriers for in situ treatment of contaminated groundwater. Pp. 881-887 in Proceedings of the International Containment Technology Conference, St. Petersburg, Fla., February 9-12.

Smyth, D., R. Joewtt, and M. Gamble. 1997b. Sealable joint steel sheet piling for groundwater control and remediation: Case histories. Pp. 206-214 in Proceedings of the International Containment Technology Conference, St. Petersburg, Fla., February 9-12.

Snyder, G., G. Mergia, and S. Cook. 1997. Containment performance assessment through hydraulic testing–Baltimore works site with comparison. Pp. 1046-1052 in Proceedings of the International Containment Technology Conference, St. Petersburg, Fla., February 9-12.

Spalding, B. P., J. S. Tixier, and C. L. Timmerman. 1997. In situ vitrification field-scale treatability study at ORNL seepage pit 1. Poster presented at Ninth National Technology Information Exchange Workshop, Shilo Inn, Idaho Falls, Idaho, August 26-28.

Steeples, D. W., and R. D. Miller. 1993. Basic principles and concepts of practical shallow seismic reflection profiling. Mining Engineering October:1297-1302.

Stumm, W. 1992. Chemistry of the Solid-Water Interface. New York: Wiley.

Stumm, W., and J. J. Morgan. 1981. Aquatic Chemistry: An Introduction Emphasizing Chemical Equilibria in Natural Waters. (Second Edition). New York: Wiley.

Subsurface Contaminants Focus Area (SCFA). 1996. Subsurface Contaminants Focus Area Technology Summary. DOE/EM-0296. Oak Ridge, Tenn.: Office of Scientific and Technical Information.

Tessier, A., P. G. C. Campbell, and M. Bisson. 1979. Sequential extraction procedure for the speciation of particulate trace metals. Analytical Chemistry 51:844-851.

Tipping, E. 1994. WHAM–A chemical equilibrium model and computer code for waters, sediments, and soils incorporating a discrete site/electrostatic model of ion-binding by humic substances. Computers and Geoscience 20:973-1023.

Tipping, E., N. B. Hethering, J. Hilton, D. W. Thompson, E. Bowels, and J. Hamilton-Taylor. 1985. Artifacts in the use of selective chemical extraction to determine distributions of metals between oxides of manganese and iron. Analytical Chemistry 57:1944-1946.

Tucker, M. D. 1996. Technical Considerations for the Implementation of Subsurface Microbial Barriers for Restoration of Ground Water at UMTRCA Sites. SAND96-1459. Albuquerque, N.M.: Sandia National Laboratories.

Turick, C. E., W. A. Apel, and N. S. Carmiol,. 1996. Isolation of hexavalent chromium reducing anaerobes from hexavalent chromium-contaminated and noncontaminated environments. Applied Microbiology and Biotechnology 44:683-688.

Van den Berg, C. M. G. 1984. Determination of the complexing capacity and conditional stability constants of complexes of copper(II) with natural organic ligands in seawater by cathodic stripping voltammetry of copper-catechol complex ions. Marine Chemistry 15:1-18.

Vermeul, V. R., S. S. Teel, J. E. Amonette, C. R. Cole, J. S. Fruchter, Y. A. Gorby, F. A. Spane, J. E. Szecsody, M. D. Williams, and S. B. Yabusaki. 1995. Geologic, Geochemical, Microbiologic, and Hydrologic Characterization at the In Situ Redox Manipulation Test Site. PNL-10633. Richland, Wash.: Pacific Northwest Laboratory.

Vidic, R. D., and F. G. Pohland. 1996. Treatment Walls. Technical Evaluation Report TE-96-01. Pittsburgh, Pa.: Ground-Water Remediation Technologies Analysis Center.

Weast, R. C., Ed. 1980. CRC Handbook of Chemistry and Physics. Boca Raton, Fla.: CRC Press, Inc.

Whang, J. M. 1996. Chemical-based barrier materials. In Assessment of Barrier Containment Technologies, R. E. Rumer and J. K. Mitchell, Eds. PB96-180583. Springfield, Va.: National Technical Information Service.

Williams, C. V., S. Dalvit Dunn, and W. E. Lowry. 1997. Tracer verification and monitoring of containment systems. Pp. 1039-1045 in Proceedings of the International Containment Technology Conference, St. Petersburg, Fla., February 9-12.

Williams, M. D., S. B. Yabusaki, C. R. Cole, and V. R. Vermeul. 1994. In situ redox manipulation field experiment: Design analysis. In In-Situ Remediation: Scientific Basis for Current and Future Technologies, G. W. Gee and N. R. Wing, Eds. Richland, Wash.: Battelle Press.

Williams, M. D., S. B. Yabusaki, C. R. Cole, and V. R. Vermeul. 1994. In-situ redox manipulation field experiment: Design analysis. In Thirty-Third Hanford Symposium on Health and the Environment: In-Situ Remediation: Scientific Basis for Current and Future Technologies, Part 1, G. W. and N. R. Wing, Eds. Columbus, Ohio: Battelle Press.

Zorpette, G. 1996. Confronting the nuclear legacy: Hanford's nuclear wasteland. Scientific American (May):88-97.

4

DNAPLs: Technologies for Characterization, Remediation, and Containment

Chlorinated hydrocarbons are among the most common pollutants in groundwater and soils at Department of Energy (DOE) sites (Riley et al., 1992), as well as other contaminated sites across the United States (Pankow and Cherry, 1996). Other types of dense nonaqueous phase liquid (DNAPL) components, including polychlorinated biphenyls (PCBs), also may be present, but chlorinated solvents are by the far most prevalent (see Table 1-4). Therefore, this chapter focuses on technologies for remediation of chlorinated solvent DNAPLs, although many of these technologies are applicable to other types of DNAPLs as well.

The conventional strategies of excavating soil and pumping and treating contaminated groundwater are generally ineffective at restoration of DNAPL-contaminated sites (NRC, 1994; Pankow and Cherry, 1996). The innovative technologies discussed in this chapter have demonstrated potential for use in remediation of DNAPL-contaminated sites. However, data for these evaluations are limited because only a few well-documented pilot tests on DNAPL sites have been reported.

THE DNAPL PROBLEM

The chlorinated organic compounds that comprise the DNAPLs common at DOE sites have low solubilities in water (Table 4-1).[1] As

[1] This discussion of the DNAPL problem and the distribution of DNAPLs is after Fountain (1998).

TABLE 4-1 Properties of Select DNAPL Components Commonly Found at DOE Sites

Compound	Aqueous Solubility (mg/liter)	Density (g/cm^3)	Vapor Pressure (mm Hg)	Henry's Law Constant (atm m^3/mol)	Absolute Viscosity (cP)
Tetrachloroethylene	150	1.6227	14	1.46×10^{-2}	0.89
Trichloroethylene	1,100	1.4642	57.8	9.9×10^{-3}	0.57
1,2-Dichloroethylene	6,260	1.2565	265	5.23×10^{-3}	0.40
Trichloroethane	4,500	1.4397	19	9.09×10^{-4}	0.12
1,2-Dichloroethane	5,500	1.235	64	9.10×10^{-4}	0.80
Carbon tetrachloride	800	1.594	90	3.02×10^{-2}	0.97
Chloroform	8,000	1.483	160	3.20×10^{-3}	0.58

NOTE: Properties are at 20°C.

SOURCES: Mueller et al., 1989; Mercer and Cohen, 1990; Montgomery, 1991.

a result, when released in the subsurface they typically do not dissolve totally in the groundwater but remain largely as a separate, nonaqueous-phase liquid (NAPL). However, solubilities of these DNAPL components are much higher than drinking water standards, so they create a persistent source of groundwater contamination as they slowly dissolve. The chlorinated solvents that comprise the most common DNAPL components are denser than water (Table 4-1) and tend to sink beneath the water table. These characteristics pose a challenge to all conventional groundwater remediation technologies (NRC, 1994; Pankow and Cherry, 1996).

In the vadose zone (the soil above the water table), DNAPL flows downward under the influence of gravity with relatively little spreading (Schwille, 1988; Pankow and Cherry, 1996). Capillary forces retain a small quantity in each pore (or fracture) through which the DNAPL flows (see Box 4-1). This fraction, which is not mobile under static conditions, is termed residual saturation. Contamination in soils above the water table therefore tends to be both laterally restricted and of relatively low saturation (saturation is defined as the fraction of the pore space filled with DNAPL). Where the water table is deep, however, as at several DOE sites, the total quantity of DNAPL retained in the vadose zone may be large.

Below the water table the distribution of DNAPLs tends to be much more irregular. Entry of DNAPL into water-filled pores requires overcoming a displacement pressure resulting from capillary forces between DNAPL and water (Box 4-1). The required entry pressure increases with decreasing grain size of the solid media in the aquifer (see Table 4-2). The downward flow of DNAPL therefore

BOX 4-1
Factors Influencing the Movement of DNAPLs Underground

When two fluids are present in one environment, the fluid having a greater affinity for a solid surface tends to spread along the surface; this fluid is termed the wetting phase. Typically, water is the wetting phase relative to both air and DNAPLs, whereas DNAPLs are wetting relative to air. Capillary forces tend to favor the entry of the wetting phase into small pores or fractures. In contrast, capillary forces will resist the entrance of the nonwetting fluid into pores filled with the wetting phase (Pankow and Cherry, 1996).

In the saturated zone (the portion of the subsurface below the water table), capillary forces resist the entry of DNAPLs into water-filled pores: the required pressure (displacement pressure) for DNAPL entry increases with decreasing grain size. This has several results, including (1) a tendency for DNAPLs to spread horizontally above finer-grained layers and thus form thin horizontal layers or lenses; (2) a tendency for DNAPLs to follow preferential pathways and thus have a highly inhomogeneous distribution; and (3) the concentration of DNAPL in larger pores.

In the vadose zone (above the water table), the presence of air in soil pores allows DNAPLs to move downward without overcoming a displacement pressure. Therefore, above the water table, DNAPLs tend not to spread laterally as readily as they do in the saturated zone.

Capillary forces and contact angle also affect the retention of DNAPLs. Capillary forces retain a small fraction of DNAPLs in every pore. The amount retained, referred to as residual saturation, is higher in the saturated zone, where water keeps the DNAPLs away from pore walls, than in the vadose zone.

may be interrupted each time the DNAPL encounters a layer with a smaller grain size than the overlying one, causing the DNAPL to flow laterally above the fine-grained layer. In such cases, a DNAPL lens may accumulate until it reaches a sufficient thickness of DNAPL

TABLE 4-2 Examples of the Entry Pressure Required for Downward Migration of Trichloroethylene (TCE) in Different Media

Medium	Required Entry Pressure (cm of TCE)
Clean sand ($K = 1 \times 10^{-2}$ cm/sec)	45
Silty sand ($K = 1 \times 10^{-4}$ cm/sec)	286
Clay ($K = 1 \times 10^{-7}$ cm/sec)	4,634
Fracture, 20-μm aperture	75
Fracture, 100-μm aperture	15
Fracture, 500-μm aperture	3

NOTE: Calculations based on TCE as the DNAPL, an interfacial tension of 34 dynes/cm, a wetting angle of 0, and a porosity of 0.35. Sand and fracture entry were calculated, respectively, from Pankow and Cherry, 1996, equations 3.17 and 11.3.

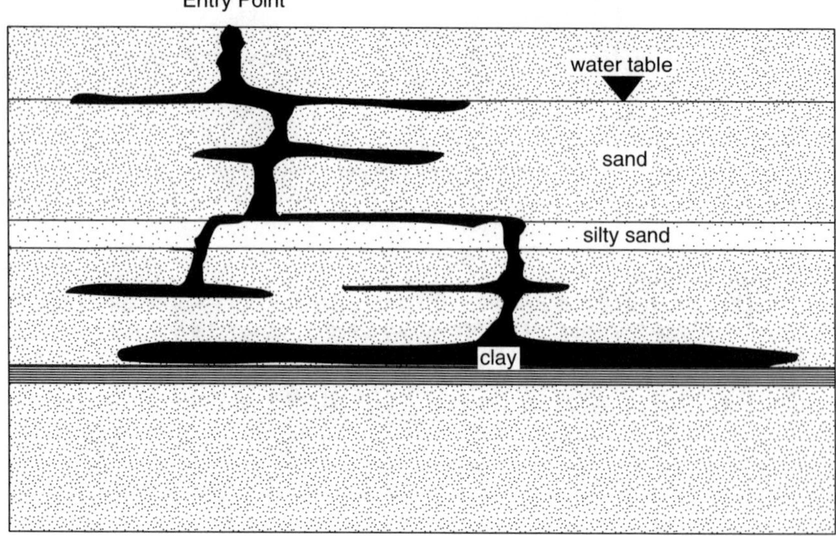

FIGURE 4-1 Typical distribution of a DNAPL in the subsurface.

to overcome the required displacement pressure. The result is a series of horizontal DNAPL lenses connected by narrow, vertical pathways (see Figure 4-1).

As in the vadose zone, a small amount of DNAPL is retained as residual saturation in every pore through which it flows. If the DNAPL encounters a layer that has a sufficiently high entry pressure, the DNAPL will be retained as a pool on the top of this layer. Thus, DNAPL is typically found in multiple horizontal lenses connected by sparse vertical pathways, with one or more pools above fine-grained layers. Most of the horizontal lenses and vertical pathways will be at or below residual saturation; only pools will have higher saturations. The distinction between residual saturation and pools is important, since only the DNAPL in pools is expected to be mobile. The flow of DNAPL and the resultant distribution are discussed in detail in Pankow and Cherry (1996).

A plume of dissolved contaminants, known as the dissolved-phase plume, will form when groundwater contacts either residual saturation, DNAPL pools, or DNAPL lenses. The volume of soil that contains DNAPL at or above residual saturation is termed the source zone. Removal of the source zone is required for restoration, because, if the source zone remains, groundwater will continue to be contaminated, typically for very long periods. Some of the remediation technologies discussed in this chapter (summarized in Tables 4-3 and 4-4) are primarily for treating source zones, whereas others are for plumes of dissolved contaminants.

CHARACTERIZATION OF DNAPL CONTAMINATION

The presence of DNAPL makes site characterization both more critical and more difficult due to the irregular distribution of DNAPLs. Site characterization includes characterization of a site's hydrogeology and of contaminant distribution at the site.[2] Because the distribution of DNAPLs is controlled by heterogeneities in the soil or rock and because the effectiveness of most technologies is also affected by heterogeneities, careful characterization of the hydrogeology of the site is essential. Technologies for hydrogeological characterization are not unique to DNAPL-contaminated sites and are thus not discussed in this report. Characterization of contaminant distribution includes defining the extent of both dissolved-phase contamination and the DNAPL source areas. Methods for defining dissolved-phase plumes are well established and are not discussed here. Determination of the limits of the DNAPL source zone is necessary both to ensure that all DNAPL is within the containment or treatment area and to minimize the volume to be treated (and hence cost). This section discusses emerging technologies for DNAPL source-zone characterization.

Characterization plans for DNAPL-contaminated sites must consider the risks inherent in penetrating a DNAPL pool. The high density and low viscosity of chlorinated solvent DNAPLs create a significant risk of mobilization if a pool is penetrated by drilling during site characterization. A DNAPL pool perched on a fine-grained layer may drain down a drill hole to lower, previously clean layers. In general, it is inadvisable to drill through DNAPL (Pankow and Cherry, 1996)

Direct-Push Technologies

Push-in tools, including large cone penetrometers, smaller units (e.g., geoprobes), and hand-held samplers driven by hammers, provide an extremely useful method of analyzing sites without drilling wells,

---

[2] For a detailed discussion of methods for characterizing DNAPL-contaminated sites, see Pankow and Cherry (1996).

TABLE 4-3 Treatment Technology Options for DNAPL Source-Zone Remediation

Technology	Mechanism(s)	Objective	Potential Application	Limitations
Steam	Volatilization, mobilization	Removal	Any aquifer unit with adequate permeability	Must be able to control steam flow and to heat units to appropriate temperatures. Heterogeneities may increase treatment time and produce tailing. Must consider implications of DNAPL mobilization.
Surfactants	Dissolution, mobilization	Removal	Any unit with adequate permeability	Must be able to establish hydraulic control. Heterogeneities may increase treatment time and produce tailing. Must consider implications of DNAPL mobilization.
Solvents	Dissolution, mobilization	Removal	Any unit with adequate permeability	Must be able to establish hydraulic control. Heterogeneities may increase treatment time and produce tailing. Must consider implications of DNAPL mobilization.
In situ oxidation	Chemical reaction	Destruction	Any unit with adequate permeability that does not react excessively with reagent	Must be able to deliver adequate reagent to source zone. Heterogeneities may increase treatment time and produce tailing. Reaction with other compounds in aquifer may reduce effectiveness.

In situ vitrification	Thermal decomposition	Destruction	Any site with appropriate depth and groundwater conditions	Soil must produce appropriate melt. Groundwater volume must not be excessive. Vapor emission must be controlled. Depth limited to about 10 m (30 ft).
Soil vapor extraction	Volatilization	Removal	Units with low soil-water content, volatile contaminants, and adequate permeability	Must be able to induce adequate air flow through entire source zone. Heterogeneities and high water contents may limit effectiveness.
Air sparging	Volatilization	Removal	Saturated units with volatile contaminants and adequate permeability	Must be able to induce air flow through entire source zone. Heterogeneities may limit effectiveness. Contaminants must be volatile at groundwater temperatures.
Electrical heating	Volatilization	Removal	Typically, low-permeability units with adequate moisture to provide conductivity	Volatilized contaminants must be removed by another technology (typically soil vapor extraction or steam). Permeability must be sufficient for vapor flow.

TABLE 4-4 Technologies Primarily for Treatment of Contaminants Dissolved from DNAPLs

Technology	Mechanism(s)	Objective	Potential Application	Limitations
In-well stripping	Volatilization	Removal	Saturated units with adequate permeability	Treatment zone depends on groundwater flow pattern established.
Bioremediation	Biologically mediated chemical reaction	Destruction	Electron acceptor, donor, and nutrients required; specific conditions depend on biodegradation pathways.	Only aqueous phase can be treated.
Reactive barriers	Chemical reaction	Dechlorination	Dissolved phase must be treatable, and water must be chemically compatible with barrier.	Treats only aqueous phase. Wall must intercept entire plume. Specific limitations depend on barrier type.
Electrokinetic systems	Electrically induced mobilization	Mobilization, coupled with another technology for destruction or removal	Low-permeablity units	Applications to DNAPL not well established. Must be coupled with technology for destruction or removal of contamination.
Physical barriers	Containment	Containment	Any unit in which barriers can be physically emplaced	Limitations vary with barrier type. If not keyed into impermeable unit at base, system will be open, and contamination will not be contained.

which is costly and invasive. Push-in tools can recover or in some cases analyze core samples of soil and samples of water taken at multiple depths during insertion. Core samples provide direct evidence of DNAPL saturation, while water samples can determine the vertical extent of contamination. A thorough characterization requires many cores, each analyzed at small vertical intervals, because of the low probability of a given core intersecting the inhomogeneously distributed DNAPL.

Many new types of push-in tools that allow direct detection of contamination, without bringing soil or water samples to the surface for analysis, have been developed in recent years. One of the most successful efforts, the site characterization and analysis penetrometer system (SCAPS) program, is discussed in detail in Chapter 5. Sensors for direct detection of contamination include the following:

- *Laser-induced fluorescence sensors.* Laser-induced fluorescence sensors use a laser to cause organic components to fluoresce in the subsurface. These tools have been highly successful in hydrocarbon detection. Although hydrocarbons, including polycyclic aromatic hydrocarbons (PAHs), generally fluoresce, most chlorinated solvents do not. Therefore this technique is of limited applicability for the detection of chlorinated solvent DNAPLs.
- *Thermal desorption volatile organic compound (VOC) sampler.* The SCAPS thermal desorption VOC sampler collects a small soil sample at a desired depth. The soil is heated, and volatile components are extracted and carried to the surface by a carrier gas. VOCs are analyzed on the surface by mass spectrometry. Multiple samples may be taken during a given push. This sampler provides essentially the same data as discrete-depth sampling of a conventional core for volatile compounds.
- *Hydrosparge VOC sensing system.* This system is mounted on a hydropunch direct-push device and pushed to the desired depth. A water sample is obtained and sparged using helium carrier gas. The gas is routed to the surface, where the sample is analyzed with an ion trap mass spectrometer.
- *Video imaging system (GeoVIS).* This system illuminates soil in contact with a sapphire window and images it with a miniature color camera. Both light nonaqueous-phase liquid (LNAPL) and DNAPL detection have been reported (Lieberman and Knowles, 1998; Lieberman et al., 1998). The NAPL phases are visible as discrete globules. Although data at this time are insufficient to establish the sensitivity of the method, the range of conditions in which it is effective, and the lower limit of NAPL saturation that can be reliably detected, the technology may offer promise for rapid source-zone delineation.

Neutron Probes

Differences in the way air and water scatter and attenuate neutrons can be used to evaluate the amount of moisture in the soil, as is done with neutron soil moisture probes. Neutron logging tools based on the difference in neutron attenuation between water and hydrocarbons are widely used to determine the saturation of oil in reservoirs. In a similar manner, the difference in attenuation of neutrons between DNAPL and water can be used to estimate DNAPL saturation. This technique is based upon the much larger neutron capture cross section of chlorine compared to water. Neutron logging tools can be used to identify intervals with DNAPL within a few centimeters of a well (Newmark et al., 1997; Daly et al., 1998).

Seismic Methods

Seismic reflection, refraction, and acoustical tomography have been used in an attempt to locate DNAPLs. Although seismic methods have been widely used for petroleum exploration and site characterization and have been highly successful in defining subsurface geology, their success in detecting DNAPLs has been limited. Generally, the differences in response due to geologic heterogeneities are greater than those due to the presence of DNAPLs. At this time, there are no well-documented successes in locating DNAPL pools using seismic methods, although research is continuing.[3]

Ground Penetrating Radar (GPR)

GPR has a proven capability to define shallow stratigraphy. It provides a detailed image of layering in the soil to a depth of from approximately a meter (a few feet) to 10 or so meters (a few tens of feet). It cannot penetrate clay layers and so is useful for defining confining layers. Where DNAPL is known to exist, GPR may have the potential to monitor its movement. However, with no prior data at a site, recognition of DNAPL cannot generally be accomplished, because the lithologic variations within a unit are generally greater than the variations caused by thin DNAPL pools (Grumman and Daniels, 1995; Pankow and Cherry, 1996; Young and Sun, 1996).

---

[3] For a review of recent developments in shallow seismic methods, see Steeples (1998).

Electrical Resistivity

DNAPLs are generally nonconductive, so methods based on conductivity or resistivity have the potential to distinguish water-filled from DNAPL-filled pores. Results to date have shown considerable success in monitoring the change in DNAPL saturation during remediation. Direct detection of DNAPL, without previous background measurements, does not generally appear to be possible (Santamarina and Fam, 1997; Daly et al., 1998; Lien and Enfield, 1998).

Partitioning Tracers

The partitioning of an organic compound between DNAPL and water is largely a function of the solubility of the compound in water. Methanol and isopropanol, which are miscible in water, partition nearly totally into the aqueous phase. On the other hand, less soluble alcohols (pentanol, heptanol, and all heavier alcohols) have limited aqueous solubilities and hence partition into DNAPL. If a solution of conservative tracers (e.g., bromide or isopropanol) and less soluble tracers (heavier alcohols) is pumped through a contaminated zone, partitioning of the less soluble compounds into the DNAPL will slow the rate of transport of the heavier alcohols exactly as sorption slows the transport of hydrophobic organic contaminants. The fraction of pore space occupied by DNAPLs can be derived from known partition coefficients and measured breakthrough curves in a well-to-well pump test (just as the retardation factor of a compound can be converted to the fraction of organic carbon in soil if the partition coefficient between soil and water is known). This technology has been demonstrated in numerous field tests (Jin et al., 1995; Brown et al., in review).

Partitioning tracers can provide a quantitative measure of the fraction of pore space occupied by DNAPLs between an injection well and an extraction well. The test averages DNAPL saturation over the entire interval. Interpretation in nonhomogeneous units requires an accurate flow model. Tracer tests require a very large number of high-quality analyses; if DNAPL saturation is low, very precise data are needed over an extended time period.

This technology has proven invaluable at several dozen sites for determining the quantity of DNAPL present and for evaluating the performance of remediation technologies in field trials through tests conducted before and after treatment. Since cleanup goals are seldom based on the amount of DNAPL left in place, but rather are based on the concentration of DNAPL components in soil or water, the utility of such tests in defining the performance of an actual remediation is less clear.

REMEDIATION TECHNOLOGIES FOR DNAPL SOURCE ZONES

Soil Vapor Extraction and Derivatives

Description

Soil vapor extraction (SVE) uses an induced flow of air through the unsaturated zone to remove volatile compounds from the soil in the vapor phase (EPA, 1997a; Holbrook et al., 1998; Johnson et al., 1993; Wilson and Clarke, 1994). In the most commonly practiced method of application, a vacuum source (e.g., a blower or vacuum pump) is connected to a well, which is screened across the contaminated interval of the unsaturated zone, as shown in Figure 4-2. The reduced pressure within the well bore induces air flow toward the well from the surrounding soils. As the air flows through the contaminated soils, the portion of volatile compounds present in the vapor phase flows toward the well and is removed through the well along with the extracted air. The volatile compounds associated with the soils (either adsorbed or dissolved in the soil moisture) or present as free-phase liquids will gradually partition into the surrounding soil gas and be extracted with the recovered air. The recovered air is either discharged directly to the atmosphere or treated and then discharged (Johnson et al., 1994). The requirements for treatment depend on the concentrations of the individual VOCs, the air flow rate, and state and local regulations.

Variants

Bioventing is similar to SVE except that the design emphasizes biodegradation rather than volatilization, with the intent to minimize physical removal (Dupont et al., 1998). Bioventing systems, like SVE systems, circulate air but require a much smaller volume of air than SVE systems. Bioventing eliminates or minimizes the need for treatment of off-gases containing volatilized contaminants. Where VOCs have no potential to enter buildings or other structures or where nonvolatile compounds are being treated, bioventing may be conducted through injection of air without air recovery. Bioventing of chlorinated compounds requires the introduction of methane, natural gas, or other substances that encourage cometabolism (fortuitous degradation of contaminants that occurs as microbes metabolize the in-

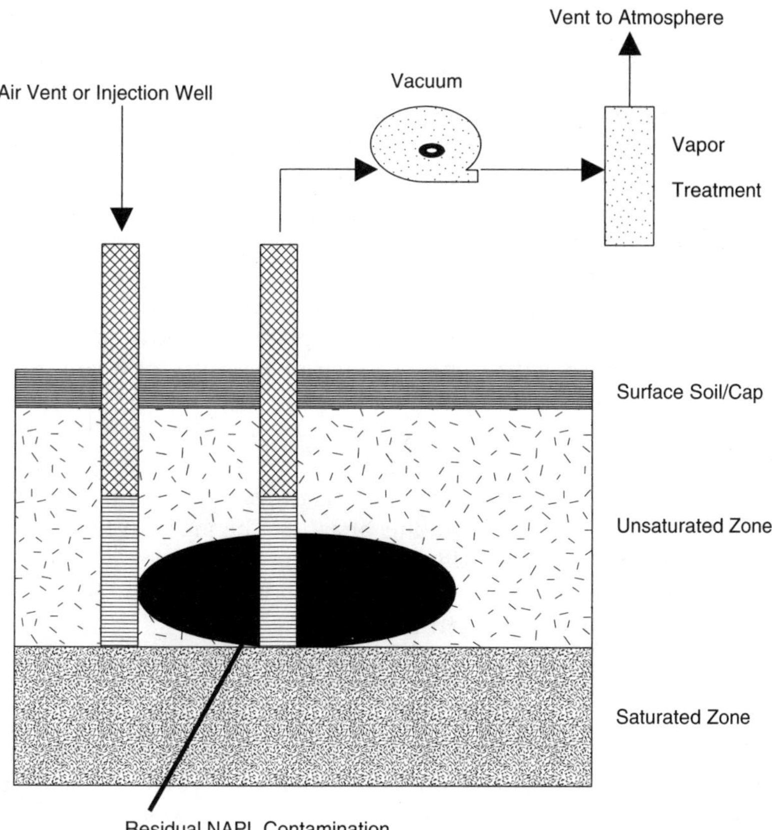

FIGURE 4-2 Typical design of an SVE system. SOURCE: Adapted from NRC, 1994.

jected compound) of the chlorinated compounds. Bioventing is discussed in more detail in the bioremediation section of this chapter.

Groundwater pumping may be used to lower the water table and thus increase the depth to which SVE may be applied. Heat may be used to increase contaminant volatility; heating may be accomplished by any of the methods discussed in the thermal technologies section of this chapter.

Physical and Chemical Principles

SVE is based on the partitioning of compounds among phases: dissolved in groundwater or air or adsorbed to soil, or present in a nonaqueous-phase liquid (NAPL) (Johnson et al., 1994; Wilson and Clarke, 1994). The concentration of a compound in the vapor phase in equilibrium with dissolved-phase contamination can be calculated from a compound's Henry's law constant. The partitioning between a NAPL and air can be calculated from a compound's vapor pressure. Due to kinetic effects, actual vapor-phase concentrations may be lower than predicted from equilibrium calculations. The vapor pressure of a compound is a function of temperature: increasing the temperature increases vapor pressure, the partitioning of a compound into the vapor phase, and thus SVE efficiency. Volatilization from nonaqueous phases and from dissolved phases is reduced if a mixture of contaminants is present. The vapor pressure of a compound in a mixture, is a function of both its pure vapor pressure and its mole fraction in the mixture as described by Raoult's law.

Because contaminants are extracted in the vapor phase, SVE performance is a function of the air movement through the soils as well as the partitioning of VOCs among phases. The amount of contaminant that can be extracted depends on the volume of air flow induced. The volume of air that SVE can induce is a function of the permeability and the water saturation of the soil.

Application

The most basic design for SVE systems uses one or more vertical wells installed by conventional drilling methods, as shown in Figure 4-2 (Johnson et al., 1994). Where there is no surface covering (such as concrete or asphalt), the tops of the well screens are located a meter or more (several feet) below the ground surface to prevent short-circuiting of air. Air is extracted from the wells in either a continuous or an intermittent mode.

Frequently, VOC recovery reaches an asymptote after an extended period of operation, making further recovery inefficient. In some instances, improved recovery rates can be achieved by operating only some of the wells at any one time using an alternating schedule. Alternatively, the system may be shut down for an extended period until a new equilibrium between the vapor phase and the combination of adsorbed, dissolved, and free-phase VOCs is established. Once the system is reactivated, recovery rates typically increase substantially.

A minor modification to the basic system is the addition of air inlet wells. These wells facilitate air entry into the subsurface, with air entering from a screened interval that matches the screened interval of the recovery wells. Wells screened in this manner can be used alternately for recovery and air inlet. Injecting air under pressure can further enhance flow through fine-grained soils.

SVE systems can use horizontal wells where the contamination and/or water table is relatively shallow or where access for installation of vertical wells is limited (e.g., beneath buildings) (Johnson et al., 1994). Horizontal wells can be installed using trenching or horizontal drilling techniques. When using trenching, horizontal trenches are excavated, a slotted pipe is placed in the trench, the pipe is covered with a porous medium such as gravel, the porous medium is covered with a low-permeability medium, and the remaining excavation is filled to the surface. Trenches can be constructed using conventional equipment, such as backhoes, or with one-pass trenching equipment. Trenches are appropriate for shallow contamination, where short-circuiting to the surface would limit the influence of vertical wells. Trenches are not appropriate where existing site infrastructure would interfere with construction. In recent years, horizontal and directional drilling equipment has permitted wells to be drilled horizontally where there is limited access to the surface above some or all of the contaminated soils. As with vertical wells, various combinations of air extraction and air injection can be used.

A recent innovation for lightly contaminated soil uses the daily changes in barometric pressure to induce air flow through wells. Wells are fitted with valves that either close or open according to the barometric pressure. During the day, when barometric pressures are relatively lower than at night, air escapes from some wells. At night, air enters a second set of wells. The result is a low flow of air into the subsurface. This type of system is more appropriate for promoting biodegradation than physical removal of VOCs because of the low air flow volumes. It is an attractive concept for remote locations.

Heating the soil can increase SVE efficiency (EPA, 1997a). The time required for remediation depends on the extent to which the contaminants partition into the vapor phase. Several methods have been evaluated for heating soils and increasing the vapor pressure of VOCs. One constraint for all potential heating methods is that soils have a large heat content. Thus, large amounts of energy are required to achieve relatively small increases in soil temperature. Methods used successfully in field-scale trials or commercial applications include radio-frequency heating, electrical resistance heating, and steam injection.

Pumping of groundwater to lower the water table may be used to increase the effective depth of SVE (EPA, 1997a). The vacuum applied to air extraction wells lowers the air pressure in the treatment zone; hence groundwater levels are slightly higher in the region of

SVE wells than would otherwise be the case. This reduces the interval over which VOCs can be removed. Extracting some groundwater from the air extraction well (dual-phase extraction) or from a nearby well maintains the groundwater at normal levels, or lower, depending on the groundwater recovery rate and soil properties. Lowering the water table by pumping increases the depth to which SVE may be applied. In aquifers with low-permeability soils, achieving significant groundwater recovery from individual wells may be difficult. For coarser-grained soils from which the water would drain fairly rapidly, large volumes of water must be recovered to lower the groundwater table sufficiently to have any benefit, and the recovered groundwater will require treatment in most cases.

One other modification that has become widely used, although developed primarily for petroleum hydrocarbons, is high-vacuum-enhanced vapor recovery. In this method, a well is screened across the water table. An annular pipe is extended into the liquid phase. The pipe and well are sealed at the surface. A high vacuum is applied, which removes water and air. In relatively low-permeability formations, this can cause dewatering and thus removal of volatile compounds from within the unsaturated and dewatered zones (Johnson et al., 1994; Kittel et al., 1994).

An important consideration in SVE is the radius of influence (ROI), the radius at which significant flow is induced for a given well. The ROI governs the well spacing needed for complete coverage of the contaminated area and hence the number of wells required. For wells with a shallow screened interval where there is no impermeable cover on the ground surface and no layer of low-permeability soils above the screened interval, air flow will have a large vertical component, with nearly all of the air flow occurring within a few feet of the well. Under these conditions, the ROI will be small. A larger ROI will be achieved when the surface is covered with concrete or asphalt, when there is a layer of low-permeability soil above the top of the well screen, and/or when the tops of the well screens are deep below the ground surface.

Design of an SVE system typically involves conducting pilot tests in representative contaminated areas, generally using a single extraction well in conjunction with several monitoring wells or probes located at several distances and at least two directions away from the extraction well (Johnson et al., 1994). Information determined includes the ROI of individual extraction wells as a function of the air extraction rate, air flow rates through soils as a function of distance from the well and air extraction rates, estimated cleanup times, off-gas treatment requirements, optimum operating conditions, and blower requirements.

SVE operations are relatively simple, consisting of maintenance of the mechanical systems, sampling of extracted air and off-gas treatment system effluent, and acquisition of occasional soil samples. Periodic review of system performance may result in modifications to operating conditions and the monitoring program. Parameters used to evaluate performance include extraction flow rates and pressures (changes may reflect changes in flow paths and ROI), vacuums, VOC concentrations, oxygen and carbon dioxide concentrations in monitoring probes, temperature, and other performance criteria associated with the mechanical components.

Performance

Generally, SVE has been most successful for treating volatile compounds in moderately to highly permeable soils. SVE systems have been applied at a large number of sites contaminated with chlorinated solvents. A general indication of SVE performance is that as of 1997, it was used at 178 sites being cleaned up under the Comprehensive Environmental Response, Compensation, and Liability Act (CERCLA) (EPA, 1998). In one DOE application, SVE in combination with six-phase electrical heating reduced perchloroethylene (PCE) concentrations by 99.7 percent in a 3-m (10-ft) clay layer. The system removed approximately 180 kg of PCE (Kittel et al., 1994). The Environmental Protection Agency (EPA) has published several case histories (EPA, 1998).

Limitations

SVE is not applicable for compounds with low volatility and Henry's law constants unless the compounds are biodegradable under aerobic conditions. It also cannot be used to treat wet, clayey soils, and is generally less applicable for remediation of low-permeability soil. The limiting effect of low pneumatic permeability is exacerbated by low vapor pressure and soil heterogeneity. To some extent, both low pneumatic permeability and limited volatility can be overcome by closely spacing wells, applying heat, or increasing the operating time of the system. These alternatives, of course, add to the cost of remediation. SVE systems in some cases can be effective for low-permeability soils if the contaminants are highly volatile and have low water solubility.

Adsorption of VOCs to soils may reduce the rate of partitioning into the soil vapor phase and interfere with SVE performance. Adsorption is important for hydrophobic compounds (those with low water solubility). The effect is proportional to the soil organic content. Adsorption to soils has the greatest effect in the later stages of remediation and can extend the time required to remove the last remnants of VOCs.

Soil heterogeneity will also affect SVE performance. Air flows most easily through coarse-grained soils and very little if at all through predominantly clayey soils. Frequently, VOCs will accumulate preferentially on the surface of and within clay lenses and layers. Air flow will then be minimal in the most highly contaminated soils. However, SVE can still decrease the contamination of the aquifer and reduce the potential for exposure to contamination if the VOCs in the less permeable soils are not very mobile.

The cost of off-gas treatment, the presence of utility trenches and other infrastructure that cause short circuiting of the extracted air, and the inability to treat metals and radionuclides are other factors that can limit use of SVE.

Advantages

SVE offers several advantages over many other remediation alternatives:

- It is an in situ technology and thus causes minimal disruption of normal site activities.
- It can be installed beneath buildings and in the vicinity of other types of infrastructure.
- It has been used at many sites and so is well developed. Design manuals are available and installation can be accomplished with readily available equipment.
- It is cost-effective for many site conditions, especially when off-gas treatment is not required or can be accomplished with existing systems, and/or for very large areas; many vendors can provide prefabricated system components.
- It is applicable to a wide variety of compounds—VOCs and aerobically biodegradable compounds.
- It will not cause further migration of contaminants.
- It reduces the potential for migration of vapors to basements and utility trenches.
- It reduces the potential for VOCs to contaminate groundwater.
- It can be used as part of a multicomponent remedial system in conjunction with a pump-and-treat system, bioremediation system, partial excavation and treatment, or natural attenuation. These methods can be implemented simultaneously with SVE or sequentially.

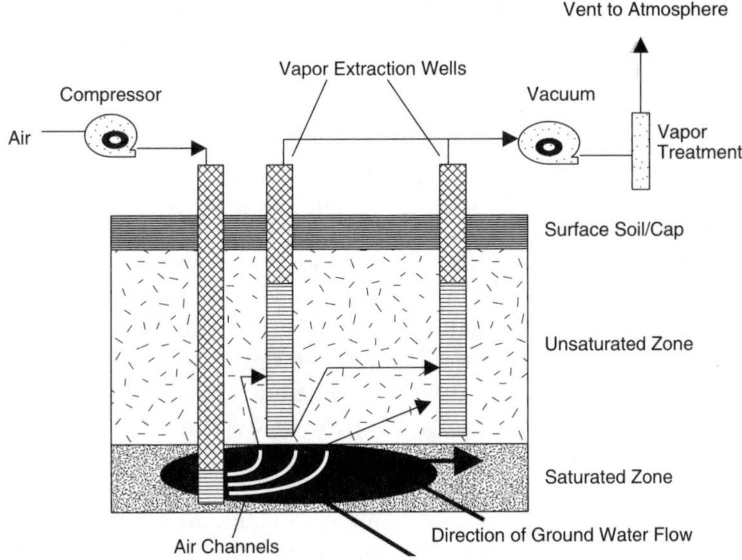

FIGURE 4-3 Typical process diagram for an air sparging system. SOURCE: Adapted from NRC, 1994.

Air Sparging

Description

Air sparging for remediation of DNAPLs involves injecting air or other gases directly into the groundwater to vaporize and recover the contaminants. Volatile components of the DNAPLs will vaporize and move upward to the atmosphere or to a vapor extraction system installed in the vadose zone (see Figure 4-3).

Physical Principles

The vaporization of volatile organic chemicals, and even mixtures of such chemicals, is well understood. Injected air moves laterally, driven by the injection pressure, and upward, due to the buoyancy of air. As the injected air moves through a formation and comes in contact with NAPLs, contaminated soil, or water containing dissolved-phase contamination, the volatile contaminants partition into the air. Partitioning from the dissolved phase is described by a compound's Henry's law constant; partitioning from DNAPLs is described by its vapor pressure. In addition, oxygen present in the injected air will dissolve in the water, promoting in situ biodegradation of nonvolatile contaminants or those located downgradient of the sparging zone.

Application

In an idealized homogeneous geological deposit, air would be injected into the saturated zone below the DNAPL and flow upward through the source zone. Since DNAPLs at a contaminated site are found in discontinuous ganglia in the saturated zone and in low spots overlying less permeable zones in or at the bottom of aquifers, achieving a uniform air flow through the entire source zone may be difficult. Air moves through saturated media by a complex process. Air must be under sufficient pressure to displace water from the medium. Once this is achieved, the medium in the immediate vicinity of an injection well becomes unsaturated. In media consisting of particles less than 1 to 2 mm in diameter (i.e., medium or finer sands), air will travel in continuous channels rather than as discrete bubbles. Thus, the air may bypass large volumes of the medium and will not directly contact DNAPLs in these areas. Injection of air below the DNAPL-contaminated zone is difficult if not impossible at sites where contaminants occur at the bottom of the aquifer.

Even small changes in permeability will have an influence on the distribution of DNAPLs and will also influence the pathways through which injected air flows. Descending DNAPLs will flow laterally along downward-sloping layers of less permeable zones until they are trapped or move further down through a more permeable feature, while air will flow laterally along upward-sloping layers of less permeable zones until it is trapped or moves up through a more permeable feature. Thus, in an inhomogeneous stratified medium, close and complete contact between a DNAPL and injected air is unlikely. However, under some unique circumstances—for example, where DNAPL is located between two less permeable confining zones that can be dewatered to a significant extent by air injected between the zones—directly vaporizing the DNAPL may be possible.

Recently, Johnson et al. (1997) conducted a carefully controlled study of the air flow distribution in a sparging system at a Navy gasoline service station in Port Hueneme, California. At this site, a large volume of gasoline leaked onto a shallow groundwater table, resulting in a 300-m source area and a 2-km dissolved plume. Sparging with single wells was undertaken at two locations at the site, one within the source area and one in the dissolved plume. Johnson et al.

conducted sparging tests at 2.4×10^{-3}, 4.7×10^{-3}, and 9.4×10^{-3} m^3/sec (5, 10, and 20 standard ft^3/min). Each increase in flow rate produced a corresponding increase in off-gas concentrations of contaminants. However, the duration of the increase was only a few days. Air distribution measurements using neutron probes and electrical resistance showed relatively sparse air distribution at 2.4×10^{-3} m^3/sec (5 ft^3/min) and a more uniform air distribution at 9.4×10^{-3} m^3/sec (20 ft^3/min). However, nearly all of the air was contained within a 3-m radius of the well (Johnson et al., 1997). Even within this radius, there were zones that received relatively little air. As a consequence, some portions of the soil within the immediate vicinity of the injection well were effectively cleaned because of direct contact with the air, whereas other portions were apparently not cleaned at all. After 18 months of operation, a number of wells showed no apparent improvement in water quality. Similar behavior was observed at the dissolved plume site. Based on this research, high air flow rates (e.g., 9.4×10^{-3} m^3/sec) are generally more effective for source-zone remediation, and close spacing of wells (e.g., 6 m) may be required to remove the bulk of the contaminants.

Performance

The majority of applications of air sparging have been for cleanup of fuel spills (Bass and Brown, 1996; Marley and Bruell, 1995). In these cases, both volatilization and enhanced biodegradation are important processes. In addition, because fuel is less dense than water, fuel source zones and groundwater plumes tend to occur near the water table. These are optimum conditions for the application of air sparging. Even so, there are relatively few documented cases in which source zones have been completely cleaned.

Air sparging also has been successful in cleaning up plumes of dissolved chlorinated solvents (Bass and Brown, 1996). In these cases, volatilization is the primary remediation mechanism. Air sparging may inhibit the anaerobic processes capable of biodegrading chlorinated solvents.

A sparging system was installed at Hill Air Force Base, Utah, to cut off a dissolved-phase trichloroethylene (TCE) plume coming from an unknown source under the runways. The injection system consisted of four wells in a line perpendicular to groundwater flow (U.S. Air Force, 1996). After about three months of operation, significant reductions in contaminant concentrations occurred at many of the monitoring points. However, the system did not achieve drinking water standards at a number of the monitoring points. As a result, the plume was not considered captured. A groundwater pump-and-treat system was installed to ensure capture of the plume.

At the Savannah River Site, DOE conducted a field demonstration of air sparging on a site contaminated with chlorinated solvents that had leaked from an unlined sediment basin. During the demonstration, the air sparging process increased the recovery of VOCs from 49.4 kg (109 lb) per day, obtained with an SVE system alone, to 58.5 kg (129 lb) per day (EPA, 1995c). Although no reports indicate that the air sparging system directly removed DNAPLs, Gordon (1998) suggested that the system may have mobilized trapped DNAPLs, because DNAPL recovery increased at an extraction well at the site.

Limitations

In order to determine if air sparging will be effective in removing a DNAPL, a very detailed investigation is required to delineate the location of the DNAPL and the heterogeneities in the air permeability of the subsurface media. The site investigation may reveal that air sparging will be effective in only a few geological deposits at the site. Even in these deposits, however, removing enough DNAPL to bring groundwater into compliance with relevant standards may not be possible. Even where reduced concentrations of contaminants in groundwater appear to indicate that air sparging has been effective, the concentrations may increase due to remaining DNAPL after air sparging is terminated. In most instances, air sparging would have to be used in conjunction with an SVE system to capture the contaminants. Air sparging will not be effective in removing contaminants with low volatilities.

Advantages

Air sparging is relatively inexpensive. In addition, because it involves the introduction of air only, rather than other substances, regulatory approval is generally straightforward. It is commercially available, and realistic cost estimates can be obtained for a given site. It has been shown to be effective at mass removal under appropriate conditions.

Alcohol or Cosolvent Flushing

Description

Alcohol or cosolvent flushing involves pumping one or more solvents, at concentrations ranging from a few to 80 percent, through

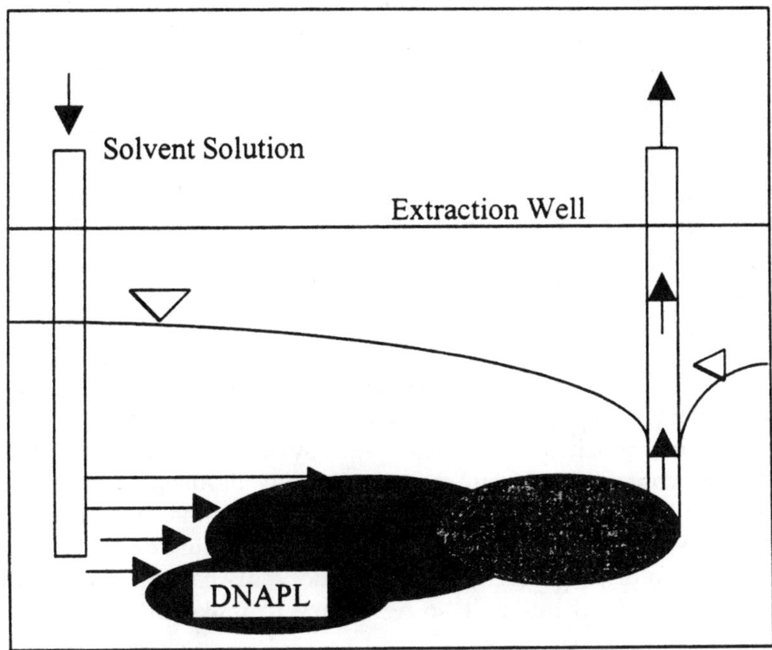

FIGURE 4-4 Process diagram for a cosolvent flushing system. SOURCE: EPA, 1995b.

the DNAPL source zone to remove DNAPL by dissolution and/or mobilization (see Figure 4-4). Alcohols are the most commonly used solvents, although in principle any organic solvent may be used. An extended discussion of the technology, including phase diagrams for several relevant systems, with interpretations, is provided in *Technology Practices Manual for Surfactants and Cosolvents* (AATDF, 1997). The manual also includes an extensive reference list.

Physical and Chemical Principles

Alcohol and cosolvent flooding relies on an increase in solubility of hydrophobic organic compounds resulting from the addition of a solvent to water and from the reduction of interfacial tension that accompanies this increased solubility. Numerous researchers have demonstrated the ability of solvents such as short-chain alcohols (methanol, ethanol, propanol) to increase the solubility of hydrophobic organic compounds in water (e.g., Rao et al., 1985; Peters and Luthy,

1993; McCray and Falta, 1996; Lunn and Kueper, 1996). Most DNAPL components found at DOE sites are readily soluble in alcohol-water mixtures. As the solubility of the DNAPL in the solvent increases (for example, by increasing the concentration of alcohol in an alcohol-water mixture), the interfacial tension between the DNAPL and water decreases. If the system is designed so that the DNAPL is miscible with the solvent flood, the interfacial tension drops to zero. Displacement of DNAPLs may occur, as well as dissolution, as the interfacial tension decreases.

Application

A typical system consists of arrays of injection and extraction wells arranged to provide an efficient flood of the source zone. Horizontal wells, trenches, or other delivery systems may be used. Either hydraulic control or containment walls may be used to contain the solvent flood. The effluent solution produced at the extraction wells contains water, solvent, and contaminants and must be treated prior to reinjection or disposal. Recycling of solvents has not been demonstrated in the field but will be necessary to make the process cost-effective.

Performance

All field trials of solvent flooding for which results have been published involved LNAPLs, which unlike DNAPLs are less dense than water. The *Technology Practices Manual for Surfactants and Cosolvents* (AATDF, 1997) and Fountain (1998) describe these trials. Three key trials occurred at Hill Air Force Base in Utah:

1. In one trial at Hill, researchers from EPA and the University of Florida tested a system that pumped approximately 10 pore volumes of a mixture of 70 percent ethanol and 12 percent pentanol in water through soil in a 3 × 5 m sheet-piling cell using a line drive array of injection and extraction wells. The contaminant treated was an LNAPL consisting of a complex mixture of weathered jet fuel and other components. The sediments within the test cell were poorly sorted sands and gravels. Initial LNAPL saturation averaged about 5 percent. The system removed approximately 85 percent of the LNAPL. Dissolution was the primary recovery mechanism.

2. In a second Hill trial, researchers from EPA and Clemson University tested a system that pumped approximately 2 pore volumes of a mixture of tertiary butanol and hexanol, followed by approxi-

mately 2.3 pore volumes of 95 percent tertiary butanol, then 0.3 pore volume of 47 percent tertiary butanol, and finally 30 pore volumes of water. The test cell, sediment, and contaminant were similar to those in the alcohol flood described above. Dissolution was the primary recovery mechanism. The system recovered between 75 and 95 percent of the LNAPL.

3. In a third test at Hill, EPA and University of Arizona researchers circulated approximately 10 pore volumes of a 10 percent cyclodextrin-in-water solution through a treatment cell with contaminants and sediments similar to those described above. Recovery ranged from 39 to 93 percent of the LNAPL originally present. Dissolution was the primary recovery mechanism.

Each of the above tests demonstrated this technology's ability to remove NAPL mass by dissolution and/or mobilization. No test achieved full recovery, although it should be noted that the Hill LNAPL is very difficult to dissolve, even in the laboratory. It is likely that these technologies would be more effective for chlorinated solvents and other easily dissolved DNAPLs such as those typically found at DOE sites.

Limitations

For any flooding technology to be effective, the entire contaminated volume of soil must be effectively flushed with treatment solutions; for solvent flooding, multiple pore volumes must be circulated. The requirement for circulation of multiple pore volumes limits application to sites with hydraulic conductivity adequate to allow large-volume pumping. The minimum conductivity will depend on the source zone size, but 10^{-4} cm/sec or greater is best in most situations.

As for all flushing technologies, heterogeneities in the aquifer will decrease extraction efficiency (Mackay and Cherry, 1989). In heterogeneous aquifers, some areas will be poorly swept by the flushing solution, and therefore such aquifers will require longer treatment times and larger treatment volumes than homogeneous aquifers. Standard numerical flow models can help predict the potential effects of heterogeneities.

Alcohols are generally less dense than water; therefore high-concentration alcohol flooding solutions will be less dense than groundwater, which sometimes presents problems in circulating these solutions evenly. Each solvent flood field trial required flooding with multiple pore volumes of treatment solution, with no recycling. Substantial volumes of solvent were used, and very large volumes of extracted fluid had to be treated. Currently, the use of large volumes of flushing solutions and the production of large volumes of extracted fluids requiring treatment are the major differences between solvent flooding and surfactant flooding (discussed below) for DNAPL remediation.

The decrease in interfacial tension produced by the addition of solvents creates the risk that DNAPLs will be mobilized. If the DNAPL is perched on an impermeable layer, lowering the interfacial tension (which reduces capillary forces and hence the entry pressure) may allow the DNAPL to penetrate the aquitard and move down into previously clean zones. The amount that the interfacial tension decreases and the risk involved depend on the hydrogeology of the site, particularly the integrity of the aquitard. This risk must be evaluated for each site.

The ultimate level to which DNAPLs may be cleaned up by cosolvent flooding is unknown, since no field trials have been reported for solvent flushing of DNAPLs. In principle, the performance of these systems should be similar to that of surfactant flooding systems.

Advantages

The chemical principles of these systems are relatively simple when treating chlorinated solvents. Alcohols are effective solvents and are not sorbed significantly. The technology is suitable for removal of DNAPLs present at very high saturations.

Surfactant-Enhanced Aquifer Remediation

Description

Remediation of DNAPL-contaminated sites with surfactants involves injection of a solution of water plus surfactant into the source zone and removal of the DNAPL through a combination of dissolution and displacement (see Figure 4-5). The relative importance of dissolution compared to displacement can be controlled by formulation of the surfactant solution.

Physical and Chemical Principles

Surfactant-enhanced remediation is based on two well-established properties of surfactants: (1) their ability to decrease interfacial tension and (2) their ability to increase the solubility of hydrophobic organic compounds. When present in sufficient concentrations (above

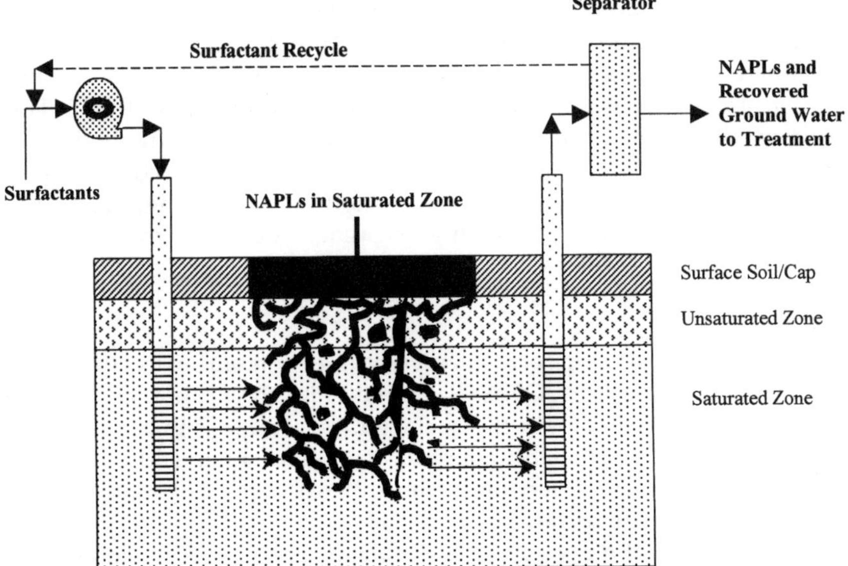

FIGURE 4-5 Typical process diagram for a surfactant flushing system. SOURCE: Adapted from NRC, 1994.

what is known as the critical micellar concentration), surfactant molecules form oriented aggregates, termed micelles. Micelles can incorporate hydrophobic molecules in their interiors, producing an apparent increase in solubility. The process of dissolving by incorporation into micelles is termed solubilization. Once solubilized, a compound is transported as if it were dissolved.

The extent of increase of solubility (solubilization) depends on the contaminant, the type of surfactant, and the surfactant concentration. Increases in solubility of more than five orders of magnitude and solubilities of hundreds of thousands of milligrams per liter have been reported for common DNAPL components (Baran et al., 1994). Early field trials used surfactants that produced modest increases in solubility (one or two orders of magnitude) and extracted the DNAPL through slow dissolution. This approach required circulation of multiple pore volumes (more than 10) of surfactant solution (Fountain et al., 1996). More recent work has emphasized higher-performance systems that would require circulation of only two to three pore volumes in a homogeneous system and would produce solubilized contaminant concentrations greater than 100,000 mg/liter (Brown et al., in review).

The interfacial tension between NAPL and water decreases as

solubilization increases. The interfacial tension may decrease by as much as four orders of magnitude at maximum solubilization. Since capillary forces decrease with decreasing interfacial tension, surfactant systems may induce DNAPL mobility (Abriola et al., 1995). This aids in recovery by increasing DNAPL flow to the extraction wells but may cause DNAPL contamination to spread if the reduced interfacial tension allows the DNAPL to penetrate underlying, previously uncontaminated layers due to high displacement pressure.

Application

A typical system involves arrays of injection and extraction wells designed to sweep the DNAPL source zone. Hydraulic controls or containment walls contain the surfactant solution. The extracted solution of water, surfactant, contaminants, and other additives must be treated prior to reinjection in the subsurface or disposal in a surface water body or sewer.

Selection of the appropriate surfactant requires consideration of performance, toxicity, biodegradability, possible chemical reactions with constituents (such as calcium) in the water, and potential for sorption. Published work has identified surfactant systems with appropriate properties that produce high solubilization of a wide range of compounds of environmental interest. Commonly, salt or a second surfactant (cosurfactant) is added to the solution to produce the desired solubilization. Alcohols may be added to optimize the phase behavior and prevent the formation of unwanted viscous phases (Lake, 1989; Pope and Wade, 1995).

After extraction, surfactants may be separated from the contaminants and reused. Air stripping has been used successfully to separate contaminants from surfactants in three field trials (at sites contaminated with PCE, carbon tetrachloride, and TCE). A permeable membrane system using solvent extraction has been developed that could be used to separate nonvolatile contaminants from the surfactant solution (AATDF, 1997). Because of the increased solubility produced by the surfactants, the extraction processes are less efficient and more costly than separation from surfactant-free water.

A modification to surfactant flooding to improve performance in heterogeneous media involves the use of mobility control agents. In surfactant floods used for enhanced oil recovery, polymers are used routinely to increase sweep efficiency (Lake, 1989). Xanthan gum, a food-grade additive, is among the most commonly used polymers. Polymers are added in low concentrations (a few hundred milligrams per liter) and produce a non-Newtonian fluid (one whose viscosity changes with flow conditions). In high-permeability units, the polymer increases the fluid's viscosity, slowing the flow. In contrast, in low-permeability layers, the high shear conditions produce a lower viscosity. Thus, the relative flow rates in low- and high-permeability zones are more nearly equal. Foam also can be used to decrease the permeability of high-permeability zones. Foam can be injected with air into a zone where surfactant has been injected (AATDF, 1997; Hirasaki et al., 1997a,b).

Performance

Surfactant systems have been field tested at more than a dozen sites, including several with DNAPLs. Representative trials, as summarized in AATDF (1997), include the following:

- Intera, Radian, and the University of Texas tested a surfactant flooding system at Hill Air Force Base in 1996. Approximately 2.5 pore volumes of a solution of surfactant, isopropanol, and sodium chloride were pumped through a poorly sorted sandy unit contaminated with a DNAPL composed primarily of TCE. A line drive system was used without confining walls. DNAPL was originally present at approximately 4 percent residual saturation. Reportedly, the system removed more than 99 percent of the DNAPL (Brown et al., in review). Concentrations in groundwater were approximately 10 mg/liter at the end of the test.
- The University of Oklahoma and the EPA also conducted a surfactant flood at Hill in 1996. Approximately 6.5 pore volumes of a mixture of 4.3 percent surfactant in water were pumped through soil in a 3 × 5 m sheet-piling cell using a line drive array of injection and extraction wells. The contaminant was an LNAPL consisting of a complex mixture of weathered jet fuel and other components. The sediments were poorly sorted sands and gravels. An estimated 90 percent of the LNAPL, which was initially present at approximately 8.5 percent saturation, was removed. The surfactant system used a mixture of two surfactants to produce a system with very low interfacial tension. The primary recovery mechanism was mobilization.
- The University of Florida and the EPA conducted a third surfactant flood in 1996. Approximately nine pore volumes of a mixture of 3 percent surfactant and 2.5 percent pentanol in water were pumped through soil within a 3 × 5 m sheet-piling cell using a line drive array of injection and extraction wells. The contaminant was an LNAPL consisting of a complex mixture of weathered jet fuel and other components. The sediments were poorly sorted sands and gravels. The surfactant system used was designed to solubilize, not mobilize the LNAPL. Soil

core data indicated that approximately 96 percent (24.9 mg/kg initial concentration reduced to 1.10 mg/kg) of the LNAPL was removed.

- The University of Buffalo and DuPont tested surfactant flood of a chlorinated solvent DNAPL in 1991-1993. A total of 12.5 pore volumes of 1 percent surfactant in water were injected using six injection wells around the treatment zone and two extraction wells near the center. The DNAPL was composed primarily of carbon tetrachloride. The contaminated unit was a sand lens within a thick clay deposit. Removal was by solubilization without mobilization. Effluent was treated by air stripping, and the surfactant solution was recycled. Cores taken at the conclusion of the test from directly between an injection and an extraction well contained no residual DNAPL. A core taken from the outer portion of the treated area contained DNAPL in the fine-grained portion but none in the higher-permeability sections. This test showed that heterogeneities limit the rate of DNAPL removal by surfactant flooding systems.

These field trials have demonstrated that surfactants can rapidly remove contaminant mass from DNAPL sites, with high removals achieved in a number of tests. However, no surfactant field tests have been continued long enough to determine the ultimate level of cleanup attainable.

Limitations

Low-permeability units, heterogeneous areas, and insoluble contaminants may impose limitations. Heterogeneities result in some portions of the treated zone receiving more solution than others, requiring a longer treatment time and larger treatment volumes than are needed for homogeneous media. Use of mobility control agents (such as polymers and foam) can minimize the effect of heterogeneities. Low-permeability formations may require very long treatment times, and circulating the required volume of surfactant solution through such formations (clays and clay-rich units) may not be practical. A hydraulic conductivity of 10^{-4} cm/sec or greater is preferred.

Although the addition of alcohol may aid in optimizing the phase behavior of the surfactant, recycling has not been demonstrated in the presence of alcohols. The very low number of pore volumes required for high-efficiency surfactants may make recycling economically unnecessary.

The use of surfactants reduces the interfacial tension between NAPL and water, thus reducing capillary forces and creating the potential for mobilization of the DNAPL. Although mobilization can be an effective technique for rapid removal of DNAPL, it also increases the risk of downward mobilization of the DNAPL. The resulting risk must be evaluated at each site based on the integrity of confining layers (aquitards) and the presence of water supplies at greater depth that could be contaminated by mobilized DNAPL.

None of the field trials at DNAPL sites has continued long enough to establish the ultimate cleanup levels that these systems can achieve in different circumstances. The persistence of some NAPL in every test conducted suggests that heterogeneities will inevitably result in some contamination remaining after treatment, although the level may be minimal and may be suitable for treatment by natural attenuation.

Advantages

Surfactant-enhanced aquifer remediation systems can rapidly remove mass from DNAPL source zones and remove DNAPLs nearly completely from relatively homogeneous units of moderate to high permeability. Many surfactants are FDA food-grade compounds and are readily biodegradable. Regulators have accepted the use of these surfactants at more than a dozen sites. Surfactant flushing can be done using conventional pumping equipment, so equipment costs are relatively low. The technology is not sensitive to operating parameters such as flow rates and concentrations. Existing numerical models can provide accurate simulations, allowing the prediction of performance for assessment purposes (Freeze et al., 1995). Implementation of the technology does not require significant site disruption, and the technology potentially could be applied beneath buildings and other structures.

In Situ Oxidation

Description

In situ oxidation systems work by injecting an oxidizing compound into the DNAPL source zone (see Figure 4-6). DNAPLs are destroyed through chemical reaction with the oxidizer. The system extracts excess oxidizer (if any) and then flushes water through the treatment zone. Potassium permanganate and hydrogen peroxide have been field tested as oxidizers in these systems.

Physical and Chemical Principles

The process is based on the ability of a strong oxidizer to destroy organic compounds. Virtually all organic contaminants can be oxi-

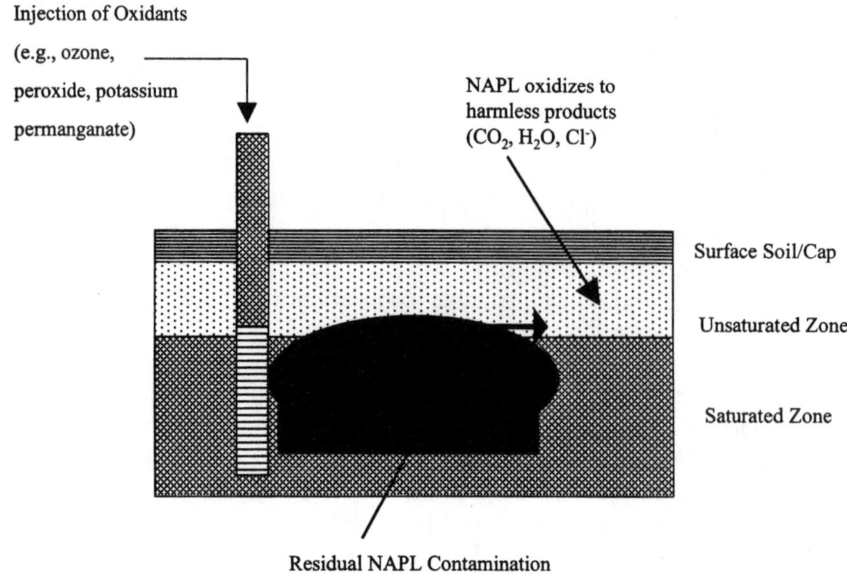

FIGURE 4-6 Process diagram for in situ oxidation. SOURCE: NRC, 1994.

dized to carbon dioxide and water under sufficiently strong oxidizing conditions. The ability of a given reagent, such as potassium permanganate or hydrogen peroxide, to oxidize a specific DNAPL can be demonstrated readily in the laboratory (e.g., Gates and Siegrist, 1995; Miller et al., 1996; Schnarr et al., 1998).

Oxidation is a nonspecific process: all compounds in the system, including solid organic matter in the soil, that can be oxidized by a given reagent will react, increasing the volume of reagent required (Miller et al., 1996). Redox reactions are often also affected by the pH of the solution, requiring acid conditions for effective oxidation in some cases. This is significant for the system of hydrogen peroxide and ferrous iron (Fenton's reagent); the reaction is optimum at low pH (2-4) and less effective at higher pH (Miller et al., 1996).

Application

The reaction of potassium permanganate or hydrogen peroxide injected in source zones (with or without ferrous iron as a catalyst) with DNAPLs yields carbon dioxide and water, plus chloride and other by-products. The extent of reaction and the end products are determined by a combination of the reagents used, the DNAPL components, and time. Potassium permanganate, or any other persistent reagent, will generally have to be washed from the treated zone by water flooding after oxidation is complete. Hydrogen peroxide spontaneously decomposes to water, with a half-life on the order of hours (Pardieck et al., 1992), so extraction of excess oxidant is not required for systems using this reagent.

Performance

A small test cell in the unconsolidated sands of Canadian Forces Base Borden was contaminated with TCE and PCE and was flushed with potassium permanganate at a concentration of 30 g/liter in a test conducted by the Solvents in Groundwater Program of the University of Waterloo. The system injected six pore volumes of permanganate, followed by clean water. VOC concentrations in water at the end of the test were near drinking water standards (Pankow and Cherry, 1996; Schnarr et al., 1998).

A field test of Fenton's reagent was conducted at DOE's Savannah River Site. In the test, 16 m^3 (4,200 gallons) of hydrogen peroxide with ferrous sulfate (to generate Fenton's reagent) were injected to a depth of 43 m (140 ft) into a saturated zone contaminated with a DNAPL consisting primarily of PCE and TCE. Researchers estimated that 94 percent of the DNAPL was destroyed in a zone of approximately 15 × 15 m (50 × 50 ft) (Jerome, 1997).

DOE conducted a relatively large-scale test using potassium permanganate at its Portsmouth, Ohio, facility in 1997. Existing horizontal wells were used to inject groundwater augmented with potassium permanganate into a sand-and-gravel zone in the X-701B area, which is contaminated with a DNAPL composed primarily of TCE. The solution was injected in one well, recovered in the other, and recirculated. A total of 780 m^3 (206 × 103 gallons) of solution was injected in a volume of approximately 67 × 27 × 1.5 m (220 × 90 × 5 ft). The volume injected corresponds to approximately 0.77 pore volume. The results, based on numerous analyses of TCE from cores taken before and after the test, indicated significant reductions in TCE in all locations reached by permanganate. Concentrations, originally as high as several hundred thousand micrograms per liter, were reduced to nondetectable levels in numerous monitoring wells immediately after the test and rebounded to low levels (tens to hundreds of micrograms per liter) after two weeks. Concentration reductions were not uniform, however. Apparently, heterogeneities in the flow field produced uneven flow of the oxidizing solution and hence uneven TCE removal. Permanganate did not reach the extraction wells during the test, so recycling of permanganate solution was not attempted (Jerome, 1997)

Limitations

Laboratory data indicate that potassium permanganate is effective for oxidation of PCE and TCE but not for destruction of chlorinated compounds without double bonds (Pankow and Cherry, 1996). In addition, strong oxidizers will react with any oxidizable compound, so this method may not be practical for treating organic-rich soils because high organic content may increase the amount of reagent required. Such soils may even react violently with strong oxidizers. The use of hydrogen peroxide, as part of Fenton's reagent (hydrogen peroxide and Fe(II) ions) works best under acidic conditions. Large amounts of calcium carbonate or other acid-soluble compounds may make maintaining the appropriate pH difficult, if not impossible. In addition, hydrogen peroxide has a limited lifetime, so it can treat only a volume that can be reached within several hours. The volume of DNAPL that can be treated also may be affected by mass transfer limitations. Because the concentration of peroxide decreases with time and because a high peroxide concentration can block subsurface pores with gas and cause the ground surface to buckle, delivering a sufficient volume of reagent to a large DNAPL pool may be difficult. The range of conditions under which this technology will be effective and the ultimate cleanup levels attainable under different scenarios have not been determined.

Because this technology requires the delivery of reagent to the entire DNAPL source zone, low-permeability zones and heterogeneities may limit performance, as they do for all flushing technologies. Specifically, since the distance a reagent penetrates from an injection well is a function of the hydraulic conductivity, in source zones with a range of conductivities a large volume of reagent or a large number of injection wells might be needed to reach lower-permeability areas within the zone.

Advantages

The initial results from Base Borden and Portsmouth suggest that potassium permanganate has considerable potential for effective destruction of PCE and TCE. Fenton's reagent has long been known to oxidize common chlorinated compounds if it can be delivered to the source zone before it degrades, and at the proper pH it is also an effective oxidizer.

Steam Injection

Description

Steam injection involves the injection of steam into a contaminated unit to volatilize and mobilize contaminants, including DNAPLs (see Figure 4-7). Condensed steam and contaminants are recovered at extraction wells. A variant of steam injection uses hot water, with the objectives of mobilizing the contaminant through reduction of viscosity and, in a commercial application termed Contained Recovery of Oily Wastes (CROW®), reducing downward migration through reduction of DNAPL density. Another variant of the process combines steam injection with direct electrical heating of fine-grained units. Since steam requires sufficient flow to supply enough heat to the entire unit, it is less effective in fine-grained units. Electrical heating may be applied to fine-grained units to drive contaminants to the steamed zones. (The use of electrical heating as a stand-alone treatment method is described in the next section.)

Physical and Chemical Principles

Steam injection promotes contaminant recovery through several mechanisms. Contaminants with boiling points lower than that of steam will volatilize. Vapor pressures of contaminants with higher boiling points will increase greatly due to the increased temperature, promoting volatilization. Finally, the increased temperatures will lower the viscosity of DNAPLs, promoting displacement (Hunt et al., 1988).

The actual process of DNAPL recovery is complex. Volatile components will enter the vapor phase and migrate away from the injection wells, toward cooler regions (Hunt et al., 1988). Condensation will occur at the thermal front, creating a bank of contaminant in front of the advancing steam. DNAPL mobilization may also occur as a result of the decreased interfacial tension and lowered velocity accompanying the increase in temperature. The relative contributions of volatilization, condensation, and displacement depend on the contaminants, site conditions, and operating parameters (Udell, 1997).

Application

Steam at the boiling point of water under the depth being treated is injected in wells, optimally bringing the entire treated volume to the boiling point of water (at the local pressure). The recovered

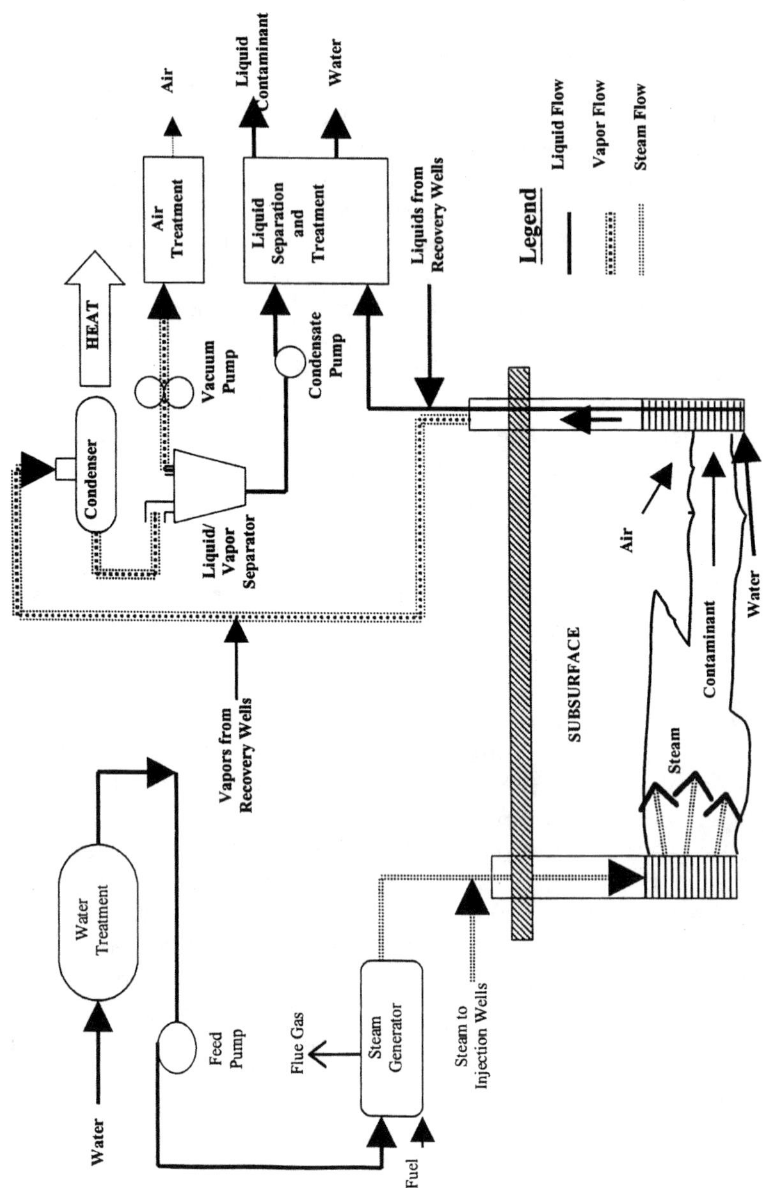

FIGURE 4-7 Process schematic of steam-enhanced extraction. SOURCE: Udell, 1997.

fluids (hot water plus contaminants) must be treated at the surface. Steam generators and steam handling equipment are commercially available.

Performance

Researchers from Lawrence Livermore National Laboratory (LLNL) and the University of California, Berkeley, conducted a combined demonstration of steam and electrical heating, called "dynamic underground stripping," at LLNL in 1992-1993. The site consisted of a sequence of sands and gravels interbedded with silt and clay units. Contamination was primarily gasoline present as an LNAPL. Researchers estimated that 25 m^3 (6,500 gallons) of gasoline were present in the treatment zone prior to the start of the test. The test was conducted at and above the water table. The water table was about 30-37 m (100-120 ft) below ground surface. Steam was injected in two zones, one at 24- to 30-m (80- to 100-ft) and one at 34- to 37-m (110- to 120-ft) depth, through six injection wells arranged in a circle on the outside of the treatment zone. Air was recovered through a central extraction well. Four electrodes were emplaced in the treatment zone, and electrical resistance heating was used to heat the fine-grained layers. The treatment zone was heated to boiling (93°C, or 200°F, at applied vacuum). The system recovered more gasoline (27 m^3, or 7,000 gallons) than was originally estimated to be present. Subsequently, regulators determined that the site met cleanup standards and closed it (DOE, 1995).

At another site, LLNL researchers, in collaboration with a private company, demonstrated the use of steam for remediation of large volumes of creosote DNAPL. To date, an area of 1.7 ha (4.3 acres) has been treated to a depth of 30 m (100 ft). In the first six weeks of operation, the system recovered 90,000 kg (200,000 lb) of NAPLs, extracted 73,000 kg (29,000 lb) in the vapor phase and burned them, captured 8,000 kg (17,500 lb) on activated carbon, and destroyed an estimated 20,600 kg (45,500 lb) by in situ decomposition. At the time of preparation of this report, the outer portions of the site were clean, and operations were continuing in the central portion (Aines, 1997).

Limitations

Fine-grained zones may require electrical heating. A risk inherent in steam flooding of DNAPL sites is that the condensed solvent front at the leading edge of the steam bank may be more mobile than the original DNAPL. Low permeability and heterogeneities will reduce the effectiveness of the process and increase the amount of contamination that remains after treatment.

Advantages

Field tests have shown that steam can effectively remove petroleum hydrocarbons; large amounts of mass have been removed relatively quickly. The limited results available to date suggest that similar performance may be achieved for DNAPLs. The thermal effects of steam, including volatilization of most common chlorinated solvent DNAPL components, will enable treatment of small-scale heterogeneities in an aquifer.

Electrical Heating

Description

A variety of electrical heating methods can be used to heat contaminated soil, with the objective of volatilizing and extracting the contaminants. As described above, heating can be coupled with SVE or steam injection. Heating methods include resistance (joule) heating, microwave heating, and radio-frequency heating (described in Figure 5-6). In each case, electrical energy is applied to the soil to produce heat. Heat increases the volatility of contaminants and may induce groundwater to boil, forming steam. Contaminants are driven out of the source zone by a combination of volatilization and thermally induced vapor-phase transport. DNAPLs will volatilize if the soil is heated to near the DNAPL boiling point and may be mobilized through a reduction of viscosity as the liquids are heated.

Physical and Chemical Principles

All heating methods rely on the increase in vapor pressure that accompanies temperature increases. As soil heats, the proportion of contaminant present in the vapor phase increases. If the temperature increases to the boiling point of water, most common DNAPL components will partition strongly into the vapor phase and can be removed through vapor extraction. Heating the vapors also increases vapor flow, as vaporization of the pore water and contaminants increases vapor pressure, promoting vapor displacement.

Application

Methods used to heat contaminated soils include the following:

- *Electrical resistance heating (joule heating).* Electrical resistance heating involves inserting electrodes in the ground and passing an alternating current through the water and soil between the electrodes. The degree of heating depends on the current and the resistance of the unit. Rocks are generally nonconductive, so most current flows through soil moisture or groundwater. The current decreases as the soil dries, decreasing conductivity. The technique is thus well suited to fine-grained soils, which typically have a high soil moisture content, although vapor transport may still limit contaminant extraction.
- *Six-phase soil heating.* Six-phase soil heating is a variant of electrical resistance heating, differing in the way the alternating current is applied to the soil. The reported advantages of six-phase heating are the more even distribution of heat due to splitting of the electrical energy into six phases and the ability to use conventional three-phase alternating current as the power source (DOE, 1995).
- *Radio-frequency heating.* Radio-frequency heating uses an electrical field created by inserting antennas into the treatment zone and exciting the soil at approved frequencies (6.68-40.68 MHz). The technology has proven capable of heating low-permeability soils to more than 150°C (Edelstein et al., 1994).

Performance

DOE demonstrated six-phase soil heating at the Savannah River Site in 1993. The target area was a 3-m-thick (10-ft-thick) clay layer at a depth of 12 m (40 ft). The primary contaminants were PCE and TCE, present at maximum concentrations of 181 and 4,529 μg/kg, respectively. Six electrodes were placed in a circle with a diameter of 9 m (30 ft). An extraction well for SVE was placed in the center of the array. The temperature was raised to 100°C in the target zone and maintained for 17 days. The system reportedly removed 99.7 percent of the contamination (DOE, 1995). IIT Research Institute demonstrated radio-frequency heating at the Rocky Mountain Arsenal in 1992. In the test, 38 m^3 (50 yd^3) of clayey soils were heated to more than 250°C. Concentrations of organochlorine pesticides decreased by 97-99 percent from initial concentrations of up to 5,000 mg/kg (EPA, 1995d).

Recently, radio-frequency heating combined with SVE was demonstrated at Kirtland Air Force Base. In this field test, 0.3 m^3 (10 ft^3) of soil contaminated with petroleum hydrocarbons was heated for 42 days. Initial treatment by SVE alone removed organic compounds such as gasoline with fewer than 12 carbon atoms per molecule but did not remove the less volatile heavier fractions. Target contaminants for the radio-frequency heating test were the heavier organics such as diesel, with between 12 and 20 carbon atoms (C_{12}-C_{20}). The maximum temperature attained was 139°C. The test system removed approximately 56 percent of the diesel-range organics in the heated volume. Initial concentrations of 2,000-4,000 mg/kg were reduced to 400-1,200 mg/kg. Apparently, the heating also stimulated biodegradation.

Limitations

Because heating technologies do not recover the contaminants themselves, they must be coupled with another technology, typically SVE, for contaminant recovery. Thermal techniques are used either when the permeability of the units to air is too low to allow adequate air flow for conventional SVE or when the vapor pressure of a contaminant is too low. Limitations to the combined technology may arise from the difficulty of fully recovering mobilized vapors. Aquifer heterogeneities may create difficulties with recovery. Compounds with lower volatility will not be effectively treated. The ultimate level of cleanup possible with these systems therefore depends on the types of heterogeneities and contaminants present at the site.

Advantages

Each of the heating technologies has proven capable of heating fine-grained soils to boiling or near-boiling temperatures. At sufficiently high temperature, volatile and semivolatile compounds will volatilize, and water will be driven off as steam. Most chlorinated solvent DNAPLs are volatile enough that heating groundwater to boiling temperatures should drive off the DNAPL as a vapor phase. Thermal methods work well in fine-grained soil, which is often difficult to treat by other methods.

In Situ Vitrification

Description

In situ vitrification (ISV) is an immobilization and destruction technology designed to treat soils and other media contaminated with organic compounds, including DNAPLs, heavy metals, and/or radioactive compounds. Soils are heated until they melt by applying an alternating electrical current between electrodes placed in the ground. Temperatures may exceed 1700°C; at these temperatures, organic compounds either volatilize or are destroyed. Once the target zone is melted, it is allowed to cool, forming a glass monolith that is relatively resistant to leaching (Dragun, 1991; Oma et al., 1994; NRC, 1996;).

ISV technology was developed to immobilize radioactive isotopes, and it has been employed primarily for this purpose. The technology is discussed in detail in Chapter 3. This section focuses on the use of ISV for destruction of DNAPLs. Although the objective of ISV for treatment of metals and radionuclides is the immobilization of contaminants within the glass monolith, the objective in treating organic compounds, including DNAPLs, is their destruction by the high temperatures produced in the process.

Physical and Chemical Principles

ISV is based on joule heating. Soils have a relatively high resistance, so inducing a flow of electrical current through them generates large amounts of heat. The process is designed to produce temperatures sufficiently high to melt the soil matrix. Temperatures of approximately 1700°C are typically generated.

Organic compounds are not stable at these temperatures. As the temperature increases, organic compounds first volatilize as their boiling points are exceeded and then thermally decompose. The products of thermal decomposition depend upon oxidation conditions (whether oxidation or pyrolysis occurs). The increased vapor pressures due to heating and the creation of a low-pressure zone in the overlying hood for the system cause some organic vapors to migrate to the hood, where they are captured and treated.

As an ISV melt is conducted, the zone of melting expands gradually. A given volume of soil thus heats gradually. As the temperature exceeds the boiling point of water (at the local pressure), water vapor will be driven off, creating a zone of higher permeability to vapors on the edges of the melt. Organic material in this zone will volatilize before being thermally destroyed. The fate of the vapor (destruction in the hotter zones or flow around the melt zone) will depend on the geometry of the soil and melt.

Application

As described in Chapter 3, in the system developed by Battelle Pacific Northwest Laboratories and licensed to Geosafe, Inc., four graphite electrodes are inserted to a shallow depth in a square pattern in the soil to be treated. An electrical current is applied to melt the contaminated soil (EPA, 1997g; Oma et al., 1994).

The extreme temperatures generated by ISV volatilize many organic compounds along the heated front before the soil melts. Organic compounds that do not volatilize are pyrolized. Volatilized compounds will move away from the melt because heating of the soil vapors increases the vapor pressure adjacent to the melt. The Geosafe, Inc. system manages the vapors by placing a hood over the melt area and applying a partial vacuum to the hood. Vapors are collected by the hood and treated. Off-gas treatment consists of a quencher, scrubber, demister, high-efficiency particulate air filter, and activated carbon. A thermal oxidizer may be incorporated downflow of the activated carbon system, depending on the contaminant(s) present in the matrix being treated (see discussion in Chapter 3 for more details on applications).

Performance

Table 4-5 shows sites at which ISV has been used to treat organic contaminants, as reported by Geosafe, Inc. (See Chapter 3 for additional sites, without organics, treated by ISV.) None of these tests included DNAPL, and no successful test involving DNAPLs has been reported.

TABLE 4-5 Sites at Which ISV Has Been Used for Organic Contaminants

Site	Contaminants	Comments
Wastech, Salt Lake City, Utah	Dioxins Pentachlorophenol Pesticides Herbicides VOCs, SVOCs	Excavated soils (6,000 tons total) were treated at a CERCLA site; all cleanup standards were achieved
Ube City, Japan	Organics Heavy metals	Tests were conducted with soil, mortar, asphalt, and drums
Private site, Spokane, Washington	PCBs	Test treated silty, sandy clay with debris (concrete, asphalt, protective clothing); 99.9999% contaminant destruction was reported

NOTE: SVOC = semivolatile organic compound.

Limitations

Limitations of ISV for treatment of DNAPLs are the same as those for treatment of metals and radionuclides, as described in Chapter 3. An additional limitation is that the application of ISV to DNAPLs has not been demonstrated. The key concern related to treating DNAPLs with ISV is the ability to contain DNAPL components as they volatilize. The ability of ISV to produce a sufficient temperature to destroy organics has been demonstrated. Containment of volatile phases will depend on site hydrogeology and system design.

Advantages

A significant advantage of ISV is that it is capable of treating complex geologic matrices containing mixtures of contaminants including organics and radionuclides in a single step, a capability that few technologies share. Another significant advantage for DOE sites is that treatment can occur without bringing radioactive materials to the surface, so the technology can decrease exposure risks and eliminate the need for transporting radioactive materials. A third advantage is that the cooled, vitrified mass can serve as a foundation for various types of construction, thus allowing a wide range of uses of the treated area.

REMEDIATION TECHNOLOGIES FOR PLUMES OF DISSOLVED DNAPL CONTAMINANTS

Treatment of plumes of contaminants dissolved from DNAPLs generally poses less of a technical challenge than treatment of undissolved DNAPLs because of the increased mobility of dissolved-phase contaminants. The treatment methods described in the remainder of this chapter apply primarily to dissolved-phase organic contaminants.

Electrokinetic Systems

Description

Electrokinetic treatment systems use a direct current electric potential, applied through electrodes inserted in the treatment zone, to induce migration of water and ions. The process mobilizes contaminants but does not destroy them. Either the contaminants must be recovered at the electrodes or the process must be coupled with an in situ contaminant treatment method. One type of electrokinetic system, the LASAGNA® process, combines electrokinetic migration and treatment of organic contaminants, possibly including DNAPLs (see Figure 5-8). The LASAGNA® method (developed by a consortium including Monsanto, Du Pont, General Electric, and DOE) uses a treatment zone between the electrodes to capture, or break down, contaminants as the electrokinetic process transports them. Electrokinetic processes have been used to treat metals and are described in detail in Chapter 3; this section focuses on their application to contaminants from DNAPLs.

Physical and Chemical Principles

Remediation by electrokinetics is based on the migration of water and ions in an electrical field. The movement of pore water under the influence of an electrical potential is termed electroosmosis, and the movement of ions is termed electromigration (Cabrera-Guzman et al., 1990; Acar et al., 1993, 1995). Both laboratory experiments and field work have demonstrated that electric fields can cause water and dissolved ions to migrate at significant velocities. The mechanism of movement of DNAPLs and of noncharged molecules is less well defined. DNAPL molecules, which are nonionic and generally nonpolar, would not be expected to migrate in an electrical field. DNAPLs themselves are typically nonconductive. DNAPL migration may be induced by a combination of osmotic pressure produced by the flow of water, changes in relative saturation due to the removal of water, and compaction of the unit due to dewatering. In addition, substantial temperature increases that occurred during field trials where DNAPLs were suspected to be present may have enhanced volatilization. The application of electrokinetics for treatment of organic contaminants is the objective of the LASAGNA® process.

Performance

The only documented field trials of electrokinetic systems to treat DNAPL employed the LASAGNA® process at DOE's Paducah Gaseous Diffusion Plant. A 3-m by 4.5-m zone of silts and clays contaminated with TCE was treated to a depth of 4.5 m. High dissolved-phase concentrations indicated that DNAPLs might be present, although the amount was not determined. An array of electrodes was operated for 120 days, during which TCE was reduced from an average concentration of 100-500 parts per million (ppm) to 1 ppm in the soil (approximately 99 percent removal). TCE concentrations in suspected DNAPL zones were reduced to 1 ppm, except for a zone at the base of the treatment volume. Since the volume of DNAPL present at the start was not well determined, the removal efficiency could not be estimated by mass balance. The contamination was captured on adsorbers placed between the electrodes (Ho et al., 1996).

Advantages

Data on the application of electrokinetics to contaminants from DNAPLs are insufficient to evaluate the technology's potential for this purpose. In theory, an advantage of the method is that it treats both organic contamination and metals. It is also suited to difficult-to-treat low-permeability zones.

Limitations

The mechanism of DNAPL migration is not understood, so the range of applications cannot be defined. The technology does not remove or destroy contaminants. No data are available on the levels of residual contamination that may be expected.

In Situ Bioremediation

Description

In situ bioremediation involves the in-place breakdown of contaminants by biologically mediated reactions. Bioremediation may involve no direct action to stimulate natural degradation (a method known as monitored natural attenuation, discussed later in this chapter), or it may involve addition of an electron acceptor (e.g., oxygen), nutrients, and/or an additional carbon source (a method known as engineered in situ bioremediation) (see Figure 4-8). Although organic contaminants can be degraded to carbon dioxide, water, and their component ions, biodegradation reactions may not run to completion.

At many DNAPL-contaminated sites, the DNAPL is composed of one or more common chlorinated solvents, while the dissolved-phase plume emanating from the DNAPL source zone often contains additional compounds that are metabolites of the chlorinated solvents. Common examples include the metabolites of PCE and TCE: dichloroethenes and vinyl chloride. (Ethene also may be produced but is of very limited environmental concern.) The corresponding metabolites from carbon tetrachloride are chloroform, methylene chloride, and chloromethane. Figure 4-9 shows an established metabolic pathway for PCE. The presence of metabolites that were typically not used at the sites is frequently taken as evidence that some biodegradation of

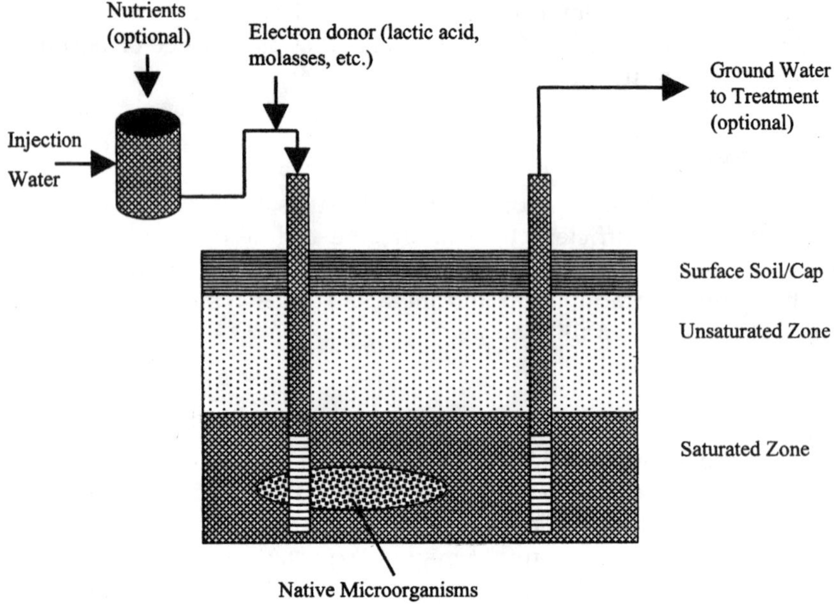

FIGURE 4-8 Process diagram for in situ bioremediation (batch or continuous addition). SOURCE: Adapted from NRC, 1994 .

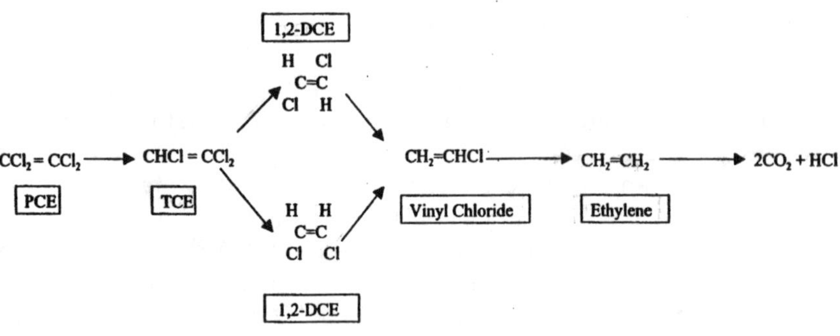

FIGURE 4-9 PCE anaerobic transformations. NOTE: DCE = dichloroethylene. SOURCE: Norris et al., 1994 .

the DNAPL components has taken place. However, the persistence of halogenated organics at many sites indicates that degradation rates are slow relative to the mass of DNAPL components typically present.

Physical, Biological, and Chemical Principles

Bioremediation of many organic contaminants is known to take place naturally in groundwater and soil under both aerobic and anaerobic conditions (Vogel et al., 1987) if environmental factors are conducive to microbial growth. Much of the recent research and development in this area has been directed toward determining which environmental factors control the rate of bioremediation and the development of methods to adjust and control these factors to increase or optimize rates. Biodegradation reactions normally involve either oxidation or reduction of the contaminant and thus require both an oxidizer (electron acceptor) and a reducer (electron donor), one of which may be the contaminant itself. The compounds serving as electron donors and electron acceptors in biodegradation reactions are referred to as primary substrates.

Many organic compounds can serve as a primary substrate; organisms use them as a carbon source and obtain energy through their metabolism. An electron acceptor is required for all such reactions. Oxygen is a common electron acceptor; nitrate, sulfate, manganese, and iron can also serve as electron acceptors in the absence of oxygen. Although aerobic degradation reactions (in which oxygen acts as the electron acceptor) are highly effective at remediation of hydrocarbons as well as some less chlorinated solvents and metabolites of chlorinated solvents, most DNAPL components resist aerobic degradation (NRC, 1993).

Most common DNAPL components, such as TCE and PCE, degrade more readily under anaerobic conditions (NRC, 1993). Although recent research suggests that chlorinated compounds may serve as primary substrates during anaerobic biodegradation under certain conditions, they often do not act as primary substrates but are degraded by cometabolic anaerobic degradation processes. In a cometabolic reaction, some other compound serves as the primary substrate. The key reaction is thought to occur between the chlorinated compound and hydrogen produced through fermentation of other organic compounds, typically degradation intermediates.

Cometabolic degradation thus requires some other carbon source to serve as a primary substrate, in addition to the compound being degraded. The carbon source may be organic carbon naturally occurring in the aquifer, a co-contaminant (petroleum hydrocarbons, for example), or a compound added by injection. This type of enhanced in situ bioremediation is in the late developmental phase. Several commercial systems have been implemented in addition to numerous field trials and demonstration projects. Organic compounds that are being tested and/or promoted by various vendors to facilitate cometabolism include benzoic acid, ethanol, propionic acid, butyric acid, lactic acid, sucrose, and molasses. Others are promoting the direct injection of hydrogen and the use of slow-release hydrogen compounds based on polymers of lactic acid.

Typically, biodegradation under anaerobic conditions is faster for more highly chlorinated compounds. As a result, *cis*-dichloroethylene (DCE) and vinyl chloride can accumulate. Reductive dechlorination of these compounds requires stronger reducing conditions. These by-products will, however, also degrade under aerobic and iron-reducing conditions that may occur at the downgradient edge of the plume. Whether these compounds accumulate therefore depends on the aquifer redox potential and other conditions. Accumulation of the degradation intermediates is a concern because these compounds are more mobile than the parent compounds (TCE and PCE) and because vinyl chloride is a carcinogen.

There is little doubt that the addition of electron donors will accelerate reductive dechlorination in aquifers where the requisite microorganisms are present. The formation of reductive dechlorination daughter products, microcosm studies with aquifer soils and groundwater, and/or speciation of microorganisms all can provide evidence that the microbes needed to carry out reductive dechlorination are present. For the process to occur, the chlorinated compounds of interest must be in solution. Thus, the process does not directly treat DNAPLs. Further, where chlorinated compounds are present in very high concentrations, by-products may be toxic to the microorganisms, thus causing the process to become self-limiting. In particular, chloride, if formed in sufficient amounts and not effectively diluted through advection, will cause osmotic shock to the organisms. Also, because chloride is generated as HCl, acid production can be problematic, although the aquifer's buffering capacity is typically sufficient to prevent large pH changes.

Because of these concerns, the use of enhanced anaerobic biodegradation is most likely to be effective in treating plumes dissolving from DNAPLs, rather than the DNAPLs themselves. Enhanced reductive dechlorination is not likely to be applicable to reducing the mass of DNAPL at most sites within a viable time period but can limit migration of dissolved-phase compounds.

Some chlorinated aliphatic compounds can degrade aerobically.

Vinyl chloride and dichloroethenes, for instance, can be used as the primary substrate by some microorganisms under both aerobic and iron-reducing conditions. Chlorinated ethenes, with the exception of PCE, can be aerobically biodegraded through a cometabolic process first described by Wilson and Wilson (1985). In this process, one of several compounds, including methane, butane, toluene, and phenol, is used as the primary substrate during aerobic degradation. Enzymes produced during this process fortuitously degrade certain chlorinated aliphatics.

Implementation of this process requires addition of both oxygen and a cometabolite. Cometabolites, along with oxygen, have been introduced in both gaseous and dissolved forms. Numerous field trials have been conducted. In general, these trials have shown that the extent of degradation of chlorinated compounds under cometabolic conditions has not been sufficient to be effective as a means of remediation.

Application

In practice, in situ bioremediation is implemented by introducing nutrients, typically nitrogen and/or phosphorus sources, air or other sources of oxygen (such as pure oxygen or hydrogen peroxide), and easily degraded organic substrates that can serve as a source of energy for the indigenous microorganisms. The types of additives will determine which microbial processes—aerobic (oxidation) or anaerobic (reductive dechlorination)—are active. In a few instances, bioaugmentation (the introduction of selected nonindigenous microorganisms into the subsurface environment to degrade specific organic contaminants) has been used (Criddle et al., 1996; Duba et al., 1996). In general, however, bioaugmentation has not provided any documented benefit.

Enhancement chemicals may be introduced through wells, infiltration galleries, or trenches, although typically wells have been used. In relatively low-permeability aquifers, multiple wells may be needed. Frequently, groundwater is recovered at downgradient locations in order to accelerate the movement of treatment chemicals through the aquifer, achieve hydraulic control, and remove some of the contaminant mass. Typically, the extracted water is treated, supplemented with enhancement chemicals, and then reintroduced into the aquifer.

Performance

In the past few years, several vendors have reported success in enhancing reductive dechlorination by adding electron donors to the subsurface. One successful implementation involved a more complex process in which sodium benzoate, sulfate, and oxygen were added at different locations within the aquifer (Beeman et al., 1994). Several vendors have claimed success in inducing reductive dechlorination by adding molasses, but little peer-reviewed information is available on field trials and commercial implementations.

A field demonstration using the combined technologies of aerobic in situ bioremediation and SVE was conducted at DOE's Savannah River Site (Brockman et al., 1995; Hazen et al., 1995; Federal Remediation Technologies Roundtable, 1997). The contaminants, including TCE and PCE, had leaked from an unlined settling basin into an aquifer consisting of several layers of sand with silt and clay beds. Concentrations of TCE ranged from 10 to 1,031 mg/liter, while PCE concentrations ranged from 3 to 124 mg/liter in the groundwater. The water table was 37 m (120 ft) below the ground surface. Nitrogen as nitrous oxide, phosphorus as triethyl phosphate, air, and methane were introduced into the aquifer via horizontal wells located 54 m (176 ft) below the ground surface. Similar wells installed 23 m (75 ft) below the ground surface were used for extraction. The injected gas moved up through the aquifer and the vadose zone to the extraction wells. During 384 days of operation, the combined systems removed 7,700 kg (17,000 lb) of volatile organic compounds and lowered the residual concentrations of TCE and PCE to below 5 mg/liter. Reportedly, bioremediation removed 40 percent more VOCs than SVE alone. Overall, TCE and PCE concentrations in the groundwater decreased by up to 95 percent. Treatment was achieved in a much shorter time period than would have been anticipated for a pump-and-treat system.

In a study conducted at Moffet Naval Air Base in California, phenol and oxygen were added to the reinjected groundwater to evaluate the ability of toluene oxygenase to serve as a cometabolic enzyme for TCE and *cis*-DCE, which were not effectively degraded when methane was used as a substrate (Hopkins et al., 1993a,b). Phenol and oxygen were added in pulses. When phenol was injected at 12 mg/liter, 85 percent of the TCE and 90 percent of the DCE were biodegraded (Hopkins et al., 1993a,b). Trichloroethane (TCA) was not significantly transformed by either of the treatments.

As a follow-up to the pilot-scale test at Moffett, a full-scale operation was established at an Edwards Air Force Base site where groundwater was severely contaminated with TCE (McCarty et al., 1998). Injection of 7 to 13.4 mg/liter of toluene, oxygen, and hydrogen peroxide into groundwater contaminated with 500 to 1,000 mg/liter of TCE effectively promoted cometabolism of the TCE. Concentrations of TCE were lowered from 1,000 mg/liter in the incoming groundwater to 18 to 24 mg/liter in the groundwater leaving the treatment zone. Removal rates for TCE and toluene were 97-98 percent and 99.98 percent, respectively.

A multiphase test was conducted at an industrial site in Watertown, Massachusetts, in conjunction with the EPA SITE (Superfund Innovative Technology Evaluation) program (Lewis et. al., 1998). A plume containing PCE, TCE, *cis*-1,2-DCE, and vinyl chloride was present in an unconfined sand and gravel aquifer. The study consisted of three phases: (1) enhanced anaerobic biodegradation using nitrogen and phosphorous sources in addition to lactic acid, (2) conversation to aerobic conditions using an oxygen release compound (ORC®), and (3) a second anaerobic phase in which hydrogen was provided using a slow-release hydrogen compound (HRC®). During the first phase, PCE was reduced from 1,500 (μ/liter to less than 100 (μ/liter, and both *cis*-1,2-DCE and vinyl chloride concentrations increased. During the second phase, *cis*-1,2-DCE and vinyl chloride concentrations decreased, and intermediate epoxides were observed. Unpublished results from the third phase indicated a 95 percent reduction in TCE and PCE concentrations and a 50 percent reduction in total mass of chlorinated ethenes.

Several reports on the application of enhanced reductive dechlorination have appeared in the non-peer-reviewed literature. For example, one report describes a site in eastern Pennsylvania, located in karst terrain 600 m (2,000 ft) upgradient of a major river (Burdick et al., 1998). Contamination resulted from metal plating wastes and sludges containing chromium and degreasing solvents. A pump-and-treat system was no longer effective in improving groundwater quality. Addition of a 2 percent molasses solution to existing wells resulted in a decrease in the oxidation-reduction potential from approximately –66 to –300 mV. A 30 percent reduction in TCE concentration, minimal increases in intermediate chlorinated ethene concentrations, and production of ethene and ethane were observed. The same authors reported a reduction of chromium to below detection limits and of TCE from 18 to 2 mg/liter at a California site.

Limitations

In situ bioremediation has several limitations. One limitation is the fact that because biodegradation occurs only in the aqueous phase, it is not suitable for direct remediation of free-phase DNAPL sources. In addition, the dechlorination of highly chlorinated hydrocarbons produces metabolites that, if not themselves degraded, are more mobile and more toxic than the original compound. In situ bioremediation also may be limited by difficulties in delivering nutrients, electron acceptors, or electron donors to low-permeability or heterogeneous zones. Therefore, some subsurface environments may not be conducive to enhanced biodegradation.

Other possible disadvantages include the potential degradation of groundwater quality by introduced nutrients and the growth of microbial biomass that may reduce the flow of water. Labor and maintenance could be costly for systems that require long-term treatment, but these costs will be incurred over significantly shorter periods than those for pump-and-treat systems.

Advantages

The major advantage of in situ bioremediation is that it uses indigenous microorganisms to treat a wide variety of soluble organic contaminants. Contaminants treated in situ are not transferred to another medium. Treatment chemicals generally move with the plume, allowing the treatment of sorbed contaminants, or they can be placed to intercept the plume. An increasing body of knowledge is available to support the systematic application of this technology in a variety of soil and geological settings. The technology can be implemented quickly, with a minimum of capital expenditure for some designs, and it is often much faster than other available options.

Phytoremediation

Description

Phytoremediation involves the use of plants to remove contaminants from soil or groundwater. As described in Chapter 3, it is an umbrella term used to describe a number of biochemical interactions that may occur among plants, microbes living on plant roots, and contaminants, and ultimately reduce contaminant concentrations (Schnoor, 1997). Potentially, it can be used to treat dissolved-phase contaminants from DNAPLs in groundwater at or near the root zone, although the use of phytoremediation for this purpose is in the very early stages of development.

Physical, Chemical, and Biological Principles

Studies have confirmed that certain plant species can take up chlorinated solvents from groundwater in the root zone (Chappell, 1997; Schnoor, 1997). Once the plant takes up the solvent, it may store the chemical in new plant structures via covalent bonding with plant lignin (Schnoor, 1997). The plant also may metabolize the chemical to other compounds. For example, Newman et al. (1997) showed that poplar trees transformed TCE to trichloroethanol, trichloroacetic acid, and dichloroacetic acid—products similar to those produced by enzymes in the human liver on exposure to TCE. Research also has indicated that the growth of plant roots can stimulate degradation of TCE by microorganisms in the root zone via aerobic cometabolism or reductive dechlorination (Chappell, 1997); the plants exude substances through their roots that can stimulate the growth of microbes required to carry out these reactions.

Application

For potential use in treating TCE contamination, phytoremediation research to date has focused primarily on using various species of poplars to serve as natural pump-and-treat systems (Chappell, 1997; Schnoor, 1997). Researchers have bred special poplars with leaves four times as large as usual to increase the rate of water and contaminant uptake. These specially bred poplars can take up and store or transform the contaminant and, in theory, provide hydraulic control of the groundwater. Poplars can extend their roots to the water table, and research studies show that a grove of poplars can create a depression in the water table ranging from several inches to several feet (Chappell, 1997). The rate at which trees pump water depends on the number of trees, tree age, time of day, season, amount of sunlight, climate, and geographic location. In studies carried out to date, pumping rates have ranged from 6 liters per day for young trees to 200 liters per day for older trees (Chappell, 1997). Schnoor (1997) provides theoretical equations for calculating groundwater capture and contaminant uptake rates to determine whether the plants can effectively control the contaminant plume.

For phytoremediation applications using hybrid poplar trees, the planting density would have to be about 1,000 to 2,000 trees per acre (Schnoor, 1997). Trees are planted from long cuttings that root and begin growing rapidly in one season. Theoretically, trees could be harvested every six years, if desired, and sold as firewood or for pulp and paper (Schnoor, 1997). If not harvested, the typical life of a poplar is 30 years.

Performance

A number of pilot-scale studies using poplars to treat TCE are under way, but final results indicating performance levels of this type of treatment system are not yet available (Chappell, 1997). At Aberdeen Proving Grounds in Maryland, 183 trees have been planted on a 1-acre site to treat TCE. At the Edward Sears Properties in New Gretna, New Jersey, the pilot test involves 118 trees planted on one-third of an acre. A third pilot test, at Carswell Air Force Base in Texas, involves 660 trees planted on 1 acre.

Limitations

Phytoremediation for treatment of dissolved chlorinated solvents is in a very early stage of development. Long-term performance cannot yet be determined for treatment of TCE because of the lack of full-scale applications. Further, phytoremediation is limited to application above the water table and in very shallow groundwater. Treating contaminants from DNAPLs that have migrated deep into groundwater will not be possible with this method.

Advantages

Phytoremediation eliminates the need for excavation and ex situ treatment. It is low cost relative to other treatment options because it is passive and solar driven (Chappell, 1997). Using poplars for TCE treatment does not generate secondary waste. Public acceptance is likely to be high because of the appeal of planting trees.

Permeable Reactive Barriers

Description

Permeable reactive barriers for groundwater remediation consist of subsurface units constructed of permeable reactive media placed to intercept the contaminated groundwater. As groundwater flows through the reactive media, dissolved contaminants are either immobilized or transformed into a more environmentally acceptable form. Removal mechanisms may include physical, chemical, and/or biological processes such as sorption, precipitation, dehalogenation, oxidation-reduction, and fixation. The selection of reactive media depends on site geochemistry, contaminant loading, required degree of contaminant concentration or mass reduction, and design lifetime of the permeable barrier.

Reactive barriers are typically envisioned as permanent or replaceable vertical walls, although horizontal applications have been considered for controlling the downward migration of contaminants.

In the case of vertical reactors, used to control laterally migrating plumes, the reactor system may extend the full width of the contaminant plume or be combined with sheet piling or low-permeability slurry walls to funnel the plume into the reactive wall. Barriers may be installed by excavating a trench and emplacing the reactive medium or by injecting reactive zones into the subsurface. In the latter case, the injection may be coupled with hydraulic fracturing techniques.

Reactive barriers may be used for treatment of water containing either, or both, dissolved organic contaminants and metals. The use of reactive barriers for the treatment of metal-contaminated water is discussed in Chapter 3. This section focuses on the use of reactive barriers to treat organic contaminants.

Numerous studies from the laboratory scale to the field scale are currently being conducted to demonstrate or improve the performance of reactive barrier technology and investigate alternative reactive media (see reviews by Shoemaker et al., 1995; EPA, 1995e, 1997f; and Vidic and Pohland, 1996). Two general classes of reactive media are currently under investigation: media that cause degradation of contaminants and media designed to sorb contaminants. The only reactive medium currently in the commercial stage of application is zero-valent iron, which is being used with funnel-and-gate technology. Most other reactive materials are in the laboratory study stage, although for some media, such as potassium permanganate grout and resting-state microorganisms, field-scale studies are being conducted to establish design and construction procedures.

Physical and Chemical Principles

Reactive barriers containing granular zero-valent iron are being used to degrade chlorinated hydrocarbons, the most common DNAPL components at DOE sites, by the process of reductive dechlorination (Gilham and O'Hannesin, 1994). The metal serves as a source of electrons for the reduction step, which removes chlorine atoms from the hydrocarbons and releases chloride and ferrous iron into solution. The reaction rate appears to be directly proportional to the surface area of granular iron present. Half-lives for the reductive dehalogenation of chlorinated hydrocarbons, normalized to 1 m^2 iron surface per milliliter of solution, range from 0.003 to 20 hours for pure iron and 0.3 to 34 hours for commercial iron (Shoemaker et al., 1995). The end products are primarily ethene and ethane, but partially dechlorinated products may form if the reaction time is insufficient. Enhancements to granular iron that result in faster degradation are being examined but have not yet been field-tested. Promising results have been obtained with palladium-plated iron (Liang et al., 1997).

Variants

Numerous researchers have evaluated variants of, or alternatives to, zero-valent iron systems. Copper and nickel salts, when added to iron filings, increase reaction rates but raise concerns about release of these metals to the aquifer. Palladium increases reaction rates without concern for additional impact on groundwater, but it is expensive. Other researchers have investigated applying a small electrical potential (e.g., 13 V) to the iron. In one reported experiment, 10.3 μM of carbon tetrachloride (CCl_4) was degraded to below the detection limit in 4 minutes compared to 3 hours without the applied voltage (Cheng and Wu, 1998). In this experiment, the pH was 7.5 and the oxidation-reduction potential was between –550 and –650 mV.

Several research groups have evaluated the use of sodium dithionite to create a permeable in situ barrier for chlorinated aliphatic hydrocarbons. Column tests conducted with aquifer materials from the Hanford Site demonstrated, like experiments conducted for hexavalent chromium reduction, that the addition of sodium dithionite could be used to create reducing conditions (Thornton et al., 1998). TCE was converted nearly completely to acetylene, with minor amounts of ethene and chloroacetylene produced. The first-order reaction was relatively slow (half-life of about 40 hours) compared to reductive dechlorination reactions with iron filings. Others found much faster degradation rates for CCl_4 (Ludwig et al., 1998) but little effect on several other chlorinated aliphatic hydrocarbons. Thornton et al. (1998) conducted tests at pH 11, while the Ludwig et al. studies were conducted at pH 7.5. This may explain the differences in reaction rates. However, increasing the aquifer pH to 11 would increase costs and probably would not be acceptable to regulatory agencies without provisions to restore the pH.

Potassium permanganate is a low-cost oxidant capable of oxidizing a wide range of organic chemicals, including chlorinated hydrocarbons. It has been commonly used in water and wastewater treatment and recently was applied successfully in a field demonstration of in situ remediation of DNAPL compounds (Siegrist et al., 1997). Potassium permanganate, a purple-colored solid crystal at room temperature, readily dissolves in water. The permanganate compound slowly decomposes to form manganese dioxide, but the degradation can be minimized by keeping the solution pH between 3 and 10. In contact with organic compounds in aqueous solution, the permanganate oxidation reactions break multiple bonds and remove the functional groups of the organic compounds. For example, double bonds in alkenes (e.g., TCE) are readily oxidized by potassium permanganate (Gates et al., 1995).

Potassium permanganate can be mixed with grout and injected or otherwise emplaced into the subsurface to form horizontal or vertical reactive barriers (Siegrist et al., 1997). The grout must be carefully chosen to ensure that it is resistant to oxidation by potassium permanganate and to provide a suitable pH for the oxidation reactions. Bentonite-based grouts appear to be the most suitable carriers for potassium permanganate.

Resting-state, indigenous microorganisms can be harnessed to create a fixed-bed biofilter, another form of reactive barrier, in which bacteria attached to aquifer material degrade chlorinated hydrocarbons (Duba et al., 1996). Microbial filters are established by biostimulation, which involves injecting electron acceptors and nutrients into the subsurface to increase the population of indigenous, contaminant-degrading microorganisms. This process is relatively simple and inexpensive: the surface operations are straightforward, and the injected compounds are generally low in cost. Creation of the biofilter may be accomplished by growing the indigenous bacteria in surface bioreactors, separating the bacteria from their growth medium, resuspending them in an aqueous solution that is devoid of added growth nutrients, and then injecting the aqueous solution into the subsurface. After the biofilter is created, ambient or induced groundwater flow delivers contaminants to the biofilter region. The degree of contaminant degradation depends on the flux of contaminants, the attached bacterial population density, and the contaminant residence time in the biofilter. Because the bacteria in the biofilter do not receive added nutrients, the performance of the filter diminishes with time, and regular replenishment of the bacterial population by injection is required.

Performance

Commercial applications and field-scale studies of permeable reactive barriers used to remediate chlorinated hydrocarbons are listed in Table 4-6, with notes regarding the treatment efficiency achieved at each site. Successful application of this technology requires an understanding of site hydrogeology and the spatial distribution of contaminants in order to determine the optimum location for the reactive barrier. Additionally, the site geochemistry must be under-

TABLE 4-6 Specifications for Commercial Applications and Field-Scale Studies of Permeable Reactive Barriers

Treatment Medium	Contaminants Treated	Demonstration Location and Date	Site and Plume Characteristics	Treatment Efficiency (VOCs)	Construction Design Notes
Granular zero-valent iron	Halogenated hydrocarbons: TCE, *cis*-1,2-DCE, VC, CFC-113	Sunnyvale, Calif.; February 1995	Semiconfined aquifer, 0.6-1.2 m	No VOCs detected within the wall	25-m-long slurry wall funnels; 12-m-long, 1.2-m-thick, 6-m-deep, 100% granular iron wall; 4-day residence time
Granular zero-valent iron	Halogenated hydrocarbons: TCE, PCE	Moffet Field, Calif.; March 1996	Shallow alluvial aquifer	TCE 16-45 μg/liter in samples taken in downgradient pea gravel section	6.5-m-long sheet pile funnels; 0.6-m pea gravel, 2-m granular iron, 0.6-m pea gravel wall totaling 3.2 m thick and 8.2 m deep; concrete beneath wall
Granular zero-valent iron	Halogenated hydrocarbons: PCE, TCE, TCA, 1,2-DCE	Coffeyville, Kans.; January 1996	800-m-long plume; sand-and-gravel unit above shale bedrock, 9 m below ground surface	Concentrations of solvents in the iron zone are below maximum contaminant levels.	150-m-long slurry wall funnels; 6-m-long, 9-m-deep 100% granular iron wall
Granular zero-valent iron	Halogenated hydrocarbons: TCE	U.S. Coast Guard Air Station, Elizabeth City, N.C.; June 1996	Shallow plume, 4-6 m below ground; water table approximately 2 m below ground	Greater than 95% reduction in chlorinated hydrocarbons	Continuous trench; 46-m-long, 7.3-m-deep, 0.6-m-thick granular iron wall
Granular zero-valent iron	Halogenated hydrocarbons: TCE, PCE	Borden, Ontario, Canada; 1991	Shallow plume 2 m wide and 1 m thick; plume was 4 m below ground and 1 m below water table	90% TCE removal and 86% PCE removal over 4-year monitoring period; 1,2-DCE detected at downgradient monitoring well	Rectangular cell (5.5 m long, 2.2 m deep, 1.5 m thick) placed 1 m deep; filled with 22% by weight granular iron and 78% by weight coarse sand

Granular zero-valent iron	Halogenated hydrocarbons: TCE, 1,2-DCE	U.S. Coast Guard Air Station, Elizabeth City, N.C.; June 1994	Shallow plume 4-6 m below ground; water table approximately 2 m below ground	~75% reduction in TCE; no change in 1,2-DCE	21 columns (20-cm diameter) installed in a staggered 3-row array to 6.7-m depth; columns filled with 50% iron filings, 25% coarse sand and 25% aquifer material
Granular zero-valent iron	Halogenated hydrocarbons: TCE, *trans*- and *cis*-1,2-DCE, VC, 1,1,-DCE,	Lowry Air Force Base, Colo.; 1995	Water table approximately 2.6 m deep; claystone bedrock 5.7 m deep	100% (all compounds below detection limits)	Funnel and gate: 3.5-m-long and 1.6-m-thick gate with 5-m-long funnel wall oriented 45 degrees of upgradient; more than 18 hours residence time
Potassium permanganate grout	Halogenated hydrocarbons: TCE	DOE-Portsmouth Gaseous Diffusion Plant, O.; 1996	Low-permeability media; highly variable amounts of DNAPLs in fractures and matrix	99% reduction within 10 cm of grout-filled fractures 3 months after emplacement	Hydraulic fracturing techniques used to emplace a permanganate-bentonite-cement grout at intervals of 0.6-1.2 m
Potassium permanganate solution	Halogenated hydrocarbons: TCE	DOE-Kansas City Plant, Kans.; 1996	Alluvial sediments, 6-10 m	Contaminant concentrations in all compliance walls are below maximum contaminant levels	Deep soil mixing with a reactive permanganate solution
Resting-state microorganisms	Halogenated hydrocarbons: TCE	Chico Municipal Airport, Calif.; 1995	Depth 28 m	98% TCE removed in initial stages until degradation capacity of bioreactor exceeded	5.4 kg (dry weight) pure strain methanotrophic bacteria injected into aquifer through single well to form quasi-spherical bioreactor with average radius 1.2 m; contaminated groundwater withdrawn through bioreactor surrounding injection well

NOTE: CFC = chlorofluorocarbon; DCA = dichloroethane; VC = vinyl chloride.

stood in order to select an appropriate reactive medium that is both sufficiently reactive to effect treatment during the time the groundwater remains in contact with the medium and sufficiently stable to be effective for an economically viable period.

Advantages

Permeable reactive barriers are a promising technology for in situ contaminant remediation. They can clean up plumes even when the source of the plume cannot be located. Because they act passively, they require no ongoing energy input and only limited maintenance after installation. Monitoring wells are generally the only surface structures visible after installation. Reactive barriers also essentially eliminate disposal requirements and disposal costs for treated waste because contaminants (except any excavated during trench installation) are not brought to the surface.

Limitations

Currently, the application of permeable reactive barriers is restricted to shallow (less than 13-m-deep), well-characterized plumes. In addition, the technology is applied mainly to dissolved contaminants. The use of reactive barriers to remediate migrating contaminant sources, such as DNAPLs, has not been tested. Reactive barriers therefore are not considered a DNAPL source-zone remediation technology. Additionally, data on the longevity of barrier reactivity and the loss of permeability due to precipitation, both subjects of significant concern, are limited.

Physical Barriers

Physical barriers, such as bentonite-slurry or sheet-piling walls, may be used to contain contamination migrating from a DNAPL source zone. Such technologies are not DNAPL remediation technologies but may be used to reduce the spread of contamination or to allow aggressive source zone remediation within the wall. These technologies are summarized in Chapter 3. Their application to DNAPL contaminants is no different from their application to sites contaminated with metals, except for the risk of DNAPL mobilization that is inherent at all DNAPL-contaminated sites. Any disturbance of a site to emplace a barrier has the potential to mobilize DNAPLs if a DNAPL pool is penetrated. Thus, accurate site characterization is required.

Natural Attenuation

Description

A variety of naturally occurring physical, chemical, and biological processes in the subsurface can decrease contaminant concentrations without human intervention. The combination of these processes is known as natural attenuation. The use of natural attenuation, with monitoring to ensure that contamination is not spreading, is becoming increasingly common for both contaminant cleanup and migration control. The EPA and state regulatory agencies have approved such monitored natural attenuation in place of or in conjunction with active remedies at a large number of sites. Monitored natural attenuation is now the leading remedy for groundwater contaminated by leaking underground storage tanks containing petroleum products (EPA, 1997d). Natural attenuation, although it is the sole remedy at only a handful of these sites, is specified as a component of the remedy in records of decision at more than one-quarter of CERCLA sites (K. Lovelace, Environmental Protection Agency, unpublished data, 1998).

Physical and Chemical Principles

The EPA, in a recent policy directive on natural attenuation, identifies the following processes as active in natural attenuation: biodegradation, biostabilization, dispersion, dilution, sorption, volatilization, and chemical transformation (see Chapter 2). For chlorinated organic contaminants, natural attenuation evaluations generally focus on biodegradation since this is almost always the primary process responsible for reducing contaminant mass. Until relatively recently, scientists believed that chlorinated organic compounds were generally highly resistant to biodegradation in the environment, but in the past two decades a variety of biological processes have been discovered that can transform these compounds in nature (for review articles, see Semprini, 1997a,b). These processes are extremely complex and not fully understood but are a topic of significant research.

The biodegradation process most frequently observed to date at sites where natural degradation of chlorinated solvents has been observed is reductive dehalogenation (Semprini, 1997a). In this process, microbes use the chlorinated compound as part of their energy metabolism, and in the process a chlorine atom is removed from the contaminant. For example, reductive dehalogenation can transform PCE, which has four chlorine atoms, to TCE, which has three, and can transform TCE to *cis*-DCE, with two chlorine atoms. *Cis*-DCE can then be reduced to vinyl chloride, which can be further reduced to ethylene (an essentially harmless compound). Buildup of some of the intermediate transformation products, especially vinyl chloride, which is more carcinogenic than the parent compounds, is a potential risk of this process. Reductive dehalogenation can occur only in anaerobic environments because it requires that the chlorinated compound serve as an electron acceptor, in place of oxygen or other electron acceptors, in microbial metabolism.

Under special conditions, some chlorinated compounds can be transformed biologically in aerobic environments (Semprini, 1997b). Aerobic transformation occurs through the process of cometabolism. In cometabolism, microorganisms do not degrade the contaminant directly, but the contaminant degrades fortuitously by enzymatic reactions that occur as the organisms metabolize other substances. Aerobic cometabolism thus requires the presence of an electron donor compound, generally methane, toluene, phenol, or some other compound that leads to production of the appropriate enzymes. The significance of aerobic cometabolism in the natural attenuation of chlorinated organic contaminant plumes is not well understood but is likely to be limited to the outer edges of the plume, where oxygen is present.

Application

Obtaining regulatory approval to use monitored natural attenuation as the sole component or as part of the remedy at a contaminated site generally requires a careful scientific study to demonstrate to regulators the extent to which natural processes are capable of controlling contaminant migration under specific conditions at the site. EPA has produced a guidance document that specifies in detail the types of evidence required at the sites it regulates (see Chapter 2 for a summary of the requirements). In addition, the Air Force has a detailed technical protocol for investigating natural attenuation of chlorinated solvents that is now widely used to guide studies of natural attenuation at non-Air Force sites (Wiedemeier et al., 1997), and EPA recently published a similar protocol based on the Air Force protocol (EPA, 1998).

Performance

A number of case studies of natural attenuation of chlorinated solvents have been conducted, some showing extensive degradation, some showing partial degradation, and some showing no degradation. One of the most extensively studied sites is a Superfund site in St. Joseph, Michigan, that is contaminated with TCE at concentrations as high as 100 mg/liter (McCarty and Wilson, 1992; Kitanidis et al., 1993; Haston et al., 1994; Wilson et al., 1994). At this site, Wilson et al. (1994) found nearly a 24-fold decrease in concentrations of chlorinated organic compounds across the site and attributed this decrease to reductive dehalogenation. Although the reductive dehalogenation was extensive, this study did not demonstrate complete transformation of the contaminants to harmless end products. The transformation products *cis*-DCE, vinyl chloride, and ethene were still present at the site. At another well-studied site, Edwards Air Force Base in California, detailed studies indicated that no biological transformation of TCE has occurred in 40 years (McCarty et al., 1998). Studies at this site have shown that the rate at which the TCE plume has grown is consistent with what would occur in the absence of biodegradation.

Limitations

Natural attenuation of chlorinated solvents is a slow process and thus will not be an appropriate strategy for sites at which relatively rapid cleanup of contaminants is required. Estimating the length of time required for transformation of the contaminants is often not possible due to the complexity of the microbial processes involved. In addition, the biological reactions responsible for attenuation of chlorinated solvents generally require the presence of other organic compounds to serve as electron donors or primary substrates; biodegradation will not occur in the absence of these other substances. Another limitation is that some transformation products that result during natural attenuation, such as vinyl chloride, are more harmful than the original contaminants and may accumulate at the site. Monitoring of sites for natural attenuation can be costly.

Advantages

The primary advantage of this method is that it can eliminate the need for an engineered solution that may disrupt the site, or it can reduce the size of the area requiring treatment with an engineered system. It also can be less costly than engineered methods, depending on the amount of site analysis and monitoring required.

COMMON LIMITATIONS OF DNAPL REMEDIATION TECHNOLOGIES

Regardless of the technology used, hydrologic and geochemical conditions will impose limitations on performance. Geological heterogeneities are the most significant cause of limitation of remediation technologies. The severity of problems produced by heterogeneities can often be predicted based on a thorough site assessment. Variation in hydraulic conductivity within the contaminated zones results in two types of problems: (1) regular lithologic variation produces channeling of flow, and (2) interunit heterogeneity results in unequal access to the unit.

Where some layers have higher conductivity than others, as is typical of layered sedimentary aquifer formations, flow will preferentially occur in the higher-conductivity units. Pumping any fluids, whether vapor or liquid, will require a longer time in the lower-conductivity units, resulting in much larger than necessary volumes being pumped through the high-permeability units.

Variations within a given unit, such as horizontal grain size variation, cause some areas to receive less flow than others within the same horizon. As a result, some zones will receive little or no treatment if a fluid is pumped in or out of the zone. In some cases, this uneven treatment is acceptable if natural attenuation rates are sufficient to control contaminants in less permeable zones following treatment of the permeable zones. The slow movement of groundwater in less permeable soils also can result in reagents being spent before they penetrate into the formation. Failure of reagents to penetrate the contaminated area is especially a problem for chemical oxidants that can oxidize naturally occurring organics and for Fenton's reagent, which can decompose to oxygen and water. In these cases, most of the reagent is consumed unproductively.

Other less obvious restrictions may result from the site lithology. For instance, sites at which a permeable, water-bearing interval is located immediately beneath a low-permeability unsaturated zone can be problematic. Remediation technologies that involve the injection of a gas phase, (e.g., air sparging, steam injection) or that can generate a gas phase (e.g., Fenton's reagent) are limited because the gas phase is difficult to collect for treatment.

Karst hydrogeology presents unique challenges. Groundwater movement in karst terrain occurs mostly through relatively large channels. Horizontal movement can be quite rapid. Reagents introduced in the source area, to the extent this can be defined, may be able to intercept the path followed by the contaminants; however, the reagent will be chasing the contaminants and may not mix sufficiently with the contaminated groundwater to produce the intended reaction.

Fractured rock poses similar challenges. Flow often occurs through a few conductive fractures. Reaching the entire source area is impossible if contaminants are located in dead-end fractures, and even defining the flow pattern is difficult.

Geochemistry varies among sites and is often affected by the release of contaminants. The groundwater pH, mineral content of soils and groundwater, and presence of nutrients beneficial to microbial processes affect many processes. For example, the release of petroleum hydrocarbons and/or oxygenated solvents, such as methyl ethyl ketone, results in the depletion of oxygen through biodegradation by the indigenous microorganisms. Following consumption of dissolved oxygen, microbial processes can result in lowered oxidation-reduction potentials and increased concentrations of reduced iron (both dissolved and on the surface of minerals), reduced manganese concentrations, and reduced sulfur species (sulfide, etc.) concentrations. Remediation processes that involve reduction, such as reduction of hexavalent chromium or reductive dechlorination of chlorinated solvents, benefit from these geochemical conditions. Conversely, these conditions will create problems for remedial technologies that introduce oxidants, resulting in excess consumption of oxidants to oxidize reduced iron, manganese, and sulfide and to increase the oxidation-reduction potential.

The challenges presented by specific hydrogeologic conditions are likely to affect all remediation technologies but are likely to be less problematic for some technologies than for others. When evaluating and selecting site remedies, it is thus necessary to investigate, understand, and consider site-specific hydrogeology and geochemistry. In some cases there will be no easy answers, and remediation, if possible at all, will be more costly and require a longer time.

CONCLUSIONS

Several technologies have shown the ability to rapidly remove mass from DNAPL source zones. Other technologies have demonstrated the ability to clean up contaminants that have dissolved from these source zones. Following are brief summaries of the demonstrated capabilities of the technologies reviewed in this chapter:

- **Soil vapor extraction** is effective for mass removal of volatile compounds in homogeneous, permeable soils and, with the addition of thermal processes, can be extended to semivolatile compounds. Removal of

DNAPLs requires sufficient flow through the entire source zone, which may be difficult to achieve.

- **Steam** can remediate DNAPLs in permeable soil in both the saturated and the unsaturated zones. It may be combined with electrical heating when fine-grained layers are present. Successful application to DNAPL remediation requires adequate permeability and control of DNAPL mobility. Heterogeneities may limit efficiency.
- **Surfactants** have demonstrated the ability to remove DNAPLs nearly completely from permeable units under saturated conditions. DNAPL remediation requires adequate permeability and consideration of DNAPL mobility. Heterogeneities may reduce efficiency.
- **Cosolvents** have shown similar potential as surfactants for rapid removal of LNAPLs and should, in principle, be equally effective with DNAPLs. DNAPL remediation requires adequate permeability and consideration of DNAPL mobility. Performance may be limited by heterogeneities.
- **In situ oxidation** has proven effective for the destruction of specific chlorinated DNAPL compounds in permeable, relatively homogeneous soils. Its application to DNAPLs requires adequate permeability and delivery of sufficient reagent to the source zone. The volume of DNAPL that may be efficiently treated may be limited by mass transfer considerations.
- **Electrical heating and electrokinetics** have shown potential for remediation of DNAPLs in low-permeability units. Both must be accompanied by some form of contaminant retrieval and destruction system. Currently, data are inadequate to determine the effectiveness of electrokinetics for remediating DNAPL source zones.
- **Biodegradation** of both chlorinated compounds and PAHs has been demonstrated. Degradation apparently takes place primarily in the dissolved phase, so bioremediation is not a direct DNAPL source-zone treatment method. Degradation of DNAPL source zones may require an extended time.
- **In situ vitrification** has demonstrated the ability to vitrify soil and produce temperatures that should lead to the destruction or mobilization of DNAPL compounds. Data on its applicability to DNAPL sites are insufficient to provide a meaningful evaluation at this time.
- **Reactive barrier walls** have shown great promise for the treatment of chlorinated solvent dissolved-phase plumes. They do not directly address the DNAPL source zone. Barrier walls, with or without reactive components, may, however, contain DNAPL source zones.

Although a range of technologies is emerging to help clean up DNAPL-contaminated sites, the number of carefully controlled field tests is insufficient to establish the ultimate cleanup level attainable for each technology. Each technology discussed in this report is based on well-established chemical, biological, and physical principles. Performance limitations are thus more likely to be a function of the hydrogeologic conditions of the site than of the processes themselves. Since an accurate characterization of the occurrence of DNAPLs is essential for the design of a remediation system and an accurate knowledge of geological heterogeneities is vital for evaluating the hydrogeological limits on remediation, thorough site characterization is required for DNAPL sites. Once site assessment has provided the means to evaluate the applicability of the technologies discussed in this report and the probable limitations of remediation, these technologies can be compared to baseline technologies such as excavation and pump-and-treat systems.

REFERENCES

AATDF (Advanced Applied Technology Demonstration Facility). 1997. Technology Practices Manual for Surfactants and Cosolvents. Houston, Tex.: AATDF, Rice University.

Abriola, L. M., K. D. Pennell, G. A. Pope, T. J. Dekker, and D. J. Luning-Prak. 1995. Impact of surfactant flushing on the solubilization and mobilization of dense nonaqueous-phase liquids. ACS Symposium 594:10-23.

Acar, Y. B., A. N. Alshawabkeh, and R. J. Gale. 1993. Fundamentals of extracting species from soils by electrokinetics. Waste Management 13:141-151.

Acar, Y. B., R. J. Gale, A. N. Alshawabkeh, R. E. Marks, S. Puppala, M. Bricka, and R. Parker. 1995. Electrokinetic remediation: Basics and technology status. Journal of Hazardous Materials 40:117-137.

Aines, R. 1997. Results from Visalia: Rapid thermal cleanup of dense nonaqueous-phase liquids. Presentation to the National Research Council Committee on Technologies for Cleanup of Subsurface Contaminants in the DOE Weapons Complex, Third Meeting, Livermore, Calif., December 15-17.

Baran, J. R. J., G. A. Pope, W. H. Wade, V. Weerasoorlya, and A. Yapa. 1994. Microemulsion formation with chlorinated hydrocarbons of differing polarity. Environmental Science and Technology 287(7):1361-1365.

Bass, D. H., and R. A. Brown. 1996. Air sparging case study database update. In Proceedings of the 1st International Symposium on In Situ Air Sparging for Site Remediation, Las Vegas, October 24-25. Potomac, Md.: INET.

Beeman, R. E., J. E. Howell, S. H. Shoemaker, E. A. Salazar, and J. R. Butram. 1994. A field evaluation of in situ microbial reductive dehalogenation by the biotransformation of chlorinated ethylenes. Pp. 14-27 in Bioremediation of Chlorinated and Polycyclic Aromatic Hydrocarbon Compounds, R.E. Hinchee, A. Leeson, L. Semprini, and S.K. Ong, Eds. Boca Raton, Fla.: Lewis Publishers.

Brockman, F. J., W. Payne, D. J. Workman, A. Soong, S. Manley, and T. C. Hazen. 1995. Effect of gaseous nitrogen and phosphorus injection on in situ bioremediation of a trichloroethylene-contaminated site. Journal of Hazardous Materials 41:287-298.

Brown, C. L., M. Delshad, V. Dwarakanath, R. E. Jackson, J. T. Londergan, H. W. Meinardus, D. C. McKinney, T. Oolman, G. A. Pope, and W. H. Wade. In review. A successful demonstration of surfactant flooding of an alluvial aquifer contaminated with DNAPL. Submitted as a research communication to Environmental Science and Technology.

Burdick, J. S., T. Bent, and S. S. Suthersan. 1998. Field applications to demonstrate natural and enhanced transformation of chlorinated aliphatic hydrocarbons. Pp. 81-86 in Natural Attenuation: Chlorinated and Recalcitrant Compounds, E. Godage, B. Wickramanayake, and R. E. Hinchee, Eds. Columbus, Oh.: Battelle Press.

Cabrera-Guzman, D., J. T. Swartzbaugh, and A. W. Weisman. 1990. The use of electrokinetics for hazardous waste site remediation. Journal of Air and Waste Management Association 40:1670-1676.

Chappell, J. 1997. Phytoremediation of TCE Using Populus. Washington, D.C.: Environmental Protection Agency, Technology Innovation Office.

Cheng, S.-C., and S.-C. Wu. 1998. Enhancing chlorinated methane degradation by modifying the Fe reduction system. Pp. 299-304 in Remediation of Chlorinated and Recalcitrant Compounds: Physical, Chemical, and Thermal Technologies, G. B. Wickramanayake and R. E. Hinchee, Eds. Columbus, Oh.: Battelle Press.

Criddle, C. S., M. Dybas, M. Witt, M. Szafranski, C. Kelly, S. Davies, M. Sneathen, S. Mathuram, J. Tiedje, L. Forney, K. Smalla, S. Bezborodnikov, L. Sepulveda-Torres, R. Brown, R. Heine, A. Chan, T. Voice, D. Wiggert, X. Zhao, O. Kawka, M. Barcelona, and T. Mayotte. 1996. The Schoolcraft field bioaugmentation experiment: Evaluation of in-situ bioaugmentation to remediate an aquifer contaminated with carbon tetrachloride. Final report submitted to State of Michigan, Department of Environmental Quality.

Daly, W., A. Ramirez, and R. Johnson. 1998. Electrical impedance tomography of a perchloroethylene release. Journal of Environmental and Engineering Geophysics 2:189-201.

DOE (Department of Energy). 1995. Dynamic Underground Stripping, Demonstrated at Lawrence Livermore National Laboratory Gasoline Spill Site: GSA, Livermore, CA. Innovative Technology Summary Report. Washington, D.C.: DOE, Office of Environmental Management and Office of Technology Development. Available at http://www.em.doe.gov/plumesfa/intech/dus/.

Dragun, J. 1991. Geochemistry and soil chemistry reactions occurring during in situ vitrification. Journal of Hazardous Materials 26:343-364.

Duba, A. G., K. J. Jackson, M. C. Jovanovich, R. B. Knapp, and R. T. Taylor. 1996. TCE remediation using in-situ resting-state bioaugmentation. Environmental Science and Technology 30:1982-1989.

Dupont, R. R., C. J. Bruell, D. C. Downey, S. G. Huling, M. C. Marley, R. D. Norris, and B. Pivetz. 1998. Innovative Site Remediation: Design and Application—Bioremediation. Annapolis, Md.: American Academy of Environmental Engineers.

Edelstein, W. A., I. E. T. Iben, O. M. Mueller, E. E. Uzgiris, H. R. Philipp, and P. B. Roemer. 1994. Radio frequency ground heating for soil remediation : Science and engineering. Environmental Progress 13(4):247-252.

EPA (Environmental Protection Agency). 1995a. Geosafe Corporation In Situ Vitrification, Innovative Technology Evaluation Report. EPA/540/R-94/520. Cincinnati, Oh.: U.S. EPA Risk Reduction Engineering Laboratory.

EPA. 1995b. In Situ Remediation Technology Status Report: Cosolvents. EPA 542-K-94-006. Washington, D.C.: EPA, Office of Solid Waste and Emergency Response.

EPA. 1995c. In situ air stripping of contaminated groundwater at U.S. Department of Energy, Savannah River Site—Aiken, South Carolina. In Remediation Case Studies: Groundwater Treatment. EPA/542/R-95/003. Washington, D.C.: EPA, Office of Solid Waste and Emergency Response.

EPA. 1995d. In Situ Remediation Technology Status Report: Thermal Enhancements. Washington, D.C.: Office of Solid Waste and Emergency Response.

EPA. 1995e. In Situ Remediation Technology Status Report: Treatment Walls. EPA 542-K-94-004. Washington, D.C.: EPA, Office of Solid Waste and Emergency Response.

EPA. 1996. Engineering Forum Issue Paper: Soil Vapor Extraction Implementation Experiences. EPA 540/F-95/030. Washington, D.C.: EPA, Office of Solid Waste and Emergency Response.

EPA. 1997a. Analysis of Selected Enhancements for Soil Vapor Extraction. EPA-542-R-97-007. Washington, D.C.: EPA, Office of Solid Waste and Emergency Response.

EPA. 1997b. Remediation Technologies Development Forum (RTDF) Update. EPA 542-F-97-005. Washington D.C.: EPA.

EPA. 1997c. Groundwater Currents, Developments in Innovative Groundwater Treatment. EPA 542-N-97-004. Washington D.C.: EPA

EPA. 1997d. Cleaning up the Nation's Waste Sites: Markets and Technology Trends. 1996 Edition. EPA 542-R-96-005. Washington, D.C.: EPA, Office of Solid Waste and Emergency Response.

EPA. 1997e. Use of Monitored Natural Attenuation at Superfund, RCRA Corrective Action, and Underground Storage Tank Sites. Directive No. 9200.4-17. Washington, D.C.: EPA, Office of Solid Waste and Emergency Response.

EPA. 1997f. Permeable Reactive Subsurface Barriers for the Interception and Remediation of Chlorinated Hydrocarbon and Chromium(VI) Plumes in Groundwater. EPA/600/F-97/008. Washington, D.C.: EPA.

EPA. 1997g. Remediation Case Studies: Bioremediation and Vitrification. Volume 5. PB97-177554. Springfield, Va.: National Technical Information Service.

EPA. 1998. Technical Protocol for Evaluating Natural Attenuation of Chlorinated Solvents in Groundwater. EPA/600/R-98/128. Washington, D.C.: EPA, Office of Research and Development.

Fatiadi, A. J. 1987. The classical permanganate ion: Still a novel oxidant in organic chemistry. Journal of Synthetic Organic Chemistry (2)85-206.

Federal Remediation Technologies Roundtable. 1997. Abstracts of Remediation Case Studies, Vol. 2. EPA 542-R-97-010. PB97-177570. Washington, D.C.: U.S. Environmental Protection Agency.

Fountain, J. C., R. C. Starr, T. Middleton, M. Beikirch, C. Taylor, and D. Hodge. 1996. A controlled field test of surfactant-enhanced aquifer remediation. Groundwater 34(5):910-916.

Fountain, J. C. 1998. Technologies for Dense Nonaqueous Phase Liquid Source Zone Remediation, Ground Water Remediation Technology Analysis Center, Technology Evaluation Report. 70 pp.

Freeze, G. A., J. C. Fountain, G. A. Pope, and R. E. Jackson. 1995. Modeling the surfactant-enhanced remediation of perchloroethylene at the Borden test site using UTCHEM compositional simulator. ACS Symposium Series #594: Surfactant-Enhanced Subsurface Remediation, D. A. Sabatini, R. C. Knox, and J. H. Harwell, Eds. Washington, D.C.: American Chemical Society.

Gates, D. D., S. R. Cline, and R. L. Siegrist. 1995. Chemical oxidation of volatile and semi-volatile organic compounds in soil. In Proceedings of the Air and Waste Management Association Conference, June.

Gates, D. D., and R. L. Siegrist. 1995. In-situ chemical oxidation of trichloroethylene using hydrogen peroxide. Journal of Environmental Engineering 121(9):639-644.

Gilham, R. W., and S. F. O'Hannesin. 1994. Enhanced degradation of halogenated aliphatics by zero-valent iron. Groundwater 32:958-967.

Gordon, M. J. 1998. Case history of a large-scale air sparging/soil vapor extraction system for remediation of chlorinated volatile organic compounds in groundwater. Groundwater Monitoring and Remediation 18:137-149.

Grumman, D., and D. Daniels. 1995. Experiments on the detection of organic contaminants in the vadose zone. Journal of Environmental and Engineering Geophysics 6:31-38.

Haston, Z. C., P. K. Sharma, N. N. Black, and P. L. McCarty. 1994. Enhanced reductive dechlorination of chlorinated ethenes. Pp. 11-14 in Symposium on Bioremediation of Hazardous Wastes: Research, Development, and Field Evaluation. EPA/600/R-94/075. Washington, D.C.: U.S. Environmental Protection Agency.

Hazen, T. C., K. H. Lombard, B. B. Looney, M. V. Enzien, J. M. Dougherty, C. B. Fliermans, J. Wear, and C. A. Eddy-Dilek. 1995. Summary of in-situ bioremediation demonstration (methane stimulation) via horizontal wells at the Savannah River Site Integrated Demonstration Project. Pp 137-150 in In-Situ Remediation: Scientific Basis for Current and Future Technologies, Part 1. G. W. Gee and N. R. Wing, Eds. Columbus, Oh.: Battelle Press.

Hirasaki, G. J., C. A. Miller, R. Szafranski, J. B. Lawson, and N. Akiya. 1997a. Surfactant/foam process for aquifer remediation. SPE 37257. Presented at Society of Petroleum Engineers International Symposium on Oilfield Chemistry, Houston, Tex., February 18-21.

Hirasaki, G. J., C. A. Miller, R. Szafranski, D. Tanzil, J. B. Lawson, H. Meinardus, M. Jin, J. T. Londergan, R. E. Jackson, G. A. Pope, and W. H. Wade. 1997b. Field Demonstration of the surfactant/foam process for aquifer remediation. SPE 39292. Presented at Society of Petroleum Engineers Technical Conference and Exhibition, San Antonio, Tex., Oct. 5-8.

Ho, Y.S., D. A. Wase, and C. G. Forster. 1996. Batch nickel removal from aqueous solution by sphagnum moss peat. Water Research 29:1327-1332.

Holbrook, T. B., D. Bass, P. Boersma, D. C. Di Giulio, J. Eisenbeis, N. J. Hutzler, and E. Roberts. 1998. Vapor Extraction and Air Sparging. Annapolis, Md.: American Academy of Environmental Engineers.

Hopkins, G. D., L. Semprini, and P. L. McCarty. 1993a. Microcosm and in-situ field studies of enhanced biotransformation of trichloroethylene by phenol-utilizing microorganisms. Applied and Environmental Microbiology 59:2277-2285.

Hopkins, G. D., J. Munakata, L. Semprini, and P. L. McCarty. 1993b. Trichloroethylene concentration effects on pilot field-scale in-situ groundwater bioremediation by phenol-oxidizing microorganisms. Environmental Science and Technology 27:2542-2547.

Hunt, J. R., N. Sitar, and K. S. Udell. 1988. Nonaqueous phase liquid transport and cleanup. 1: Analysis of mechanisms. Water Resources Research 24(8):1247-1258.

Jerome, K. 1997. In situ oxidation destruction of DNAPL. In EPA Groundwater Currents. Developments in Innovative Groundwater Treatment. EPA 542-N-97-004. Washington D.C.: U.S. Environmental Protection Agency.

Jin, M., M. Delshad, V. Dwarakanath, D. C. McKinney, G. A. Pope, K. Sepernoori, C. E. Tilburg, and R. E. Jackson. 1995. Partitioning tracer test for detection, estimation, and remediation performance assessment of subsurface nonaqueous phase liquids. Water Resources Research 31(5):1201-1211.

Johnson, P. C., A. Baehr, R. A. Brown, R. Hinchee, and G. Hoag. 1994. Innovative Site Remediation: Design and Application—Vacuum Vapor Extraction. Annapolis, Md.: American Academy of Environmental Engineers.

Johnson, P. C., R. L. Johnson, C. Neaville, E. E. Hansen, S. M. Stearns, and I. J. Dortch. 1997. An assessment of conventional in situ air sparging pilot tests. Groundwater 35(5):765-774.

Johnson, R. L., P. C. Johnson, A. Leeson, and C. M. Vogel. 1997. Air distribution during in situ air sparging: Tracer and geophysical measurements. In In Situ and On-Site Bioremediation, Vol. 1. Columbus, Oh.: Battelle Press.

Johnson, R. L., D. Johnson, D. McWhorter, R. Hinchee, and I. Goodman. 1993. An overview of in situ air sparging. Groundwater Monitoring and Remediation 13(4):127-135.

Kitanidis, P. K., L. Semprini, D. H. Kampbell, and J. T. Wilson. 1993. Natural anaerobic bioremediation of TCE at the St. Joseph, Michigan, Superfund site. Pp. 47-50 in Symposium on Bioremediation of Hazardous Wastes: Research, Development, and Field Evaluations. Cincinnati, Ohio: U.S. Environmental Protection Agency.

Kittel, J. A., R. E. Hinchee, R. Hoeppel, and R. Miller. 1994. Bioslurping: Vacuum-enhanced free product recovery coupled with bioventing—A case study. Presented at Petroleum Hydrocarbons and Organic Chemicals in Groundwater, Houston, Tex., November 2-4.

Lake, L. W. 1989. Enhanced Oil Recovery. Englewood Cliffs, N.J.: Prentice Hall.

Lewis, R. F., M. A. Dooley, J. C. Johnson, and W. A. Murray. 1998. Sequential anaerobic/aerobic biodegradation of chlorinated solvents: Pilot-scale field demonstration. Pp. 1-6 in Designing and Applying Treatment Technologies: Remediation of Chlorinated and Recalcitrant Compounds, E. Godage, B. Wickramanayake, and R. E. Hinchee, Eds. Columbus, Ohio: Battelle Press.

Liang, L., N. Korte, J. D. Goodlaxson, J. Clausen, Q. Fernando, and R. Muftikian. 1997. Byproduct formation during the reduction of TCE by zero-valence iron and palladized iron. Groundwater Monitoring and Remediation 17(1):122-127.

Lieberman, S. H., and D. S. Knowles. 1998. Cone penetrometer-deployable in situ video microscope for characterizing sub-surface soil properties. Field Analytical Chemistry and Technology 2(2):127-132.

Lieberman, S. H., G. W. Anderson, and V. Games. 1998. Use of a cone penetrometer deployed video-imaging system for in situ detection of NAPLs in subsurface soil environments. In Proceedings 1998 Petroleum Hydrocarbons and Organic Chemicals in Groundwater: Prevention, Detection and Remediation, Houston, Tex., November 11-13. Westerville, Oh.: National Groundwater Association.

Lien, B. K., and C. G. Enfield. 1998. Delineation of subsurface hydrocarbon contamination distribution using a direct push resistivity method. Journal of Environmental and Engineering Geophysics 2:173-179.

Ludwig, R., A. Rzeczkowska, and S. Dworatzek. 1998. Abiotic trichloroethylene dehalogenation using sodium hydrosulfite: Laboratory and field investigation. Pp. 347-352 in Physical, Chemical, and Thermal Technologies: Remediation of Chlorinated and Recalcitrant Compounds, G. B. Wickramanayake and R. E. Hinchee, Eds. Columbus, Oh.: Battelle Press.

Lunn, S. D., and B. H. Kueper. 1996. Removal of DNAPL pools using upward gradient ethanol floods. In Proceedings of NAPLs in the Subsurface Environment: Assessment and Remediation. Washington D.C.: American Society of Civil Engineers.

Mackay, D. M., and J. A. Cherry. 1989. Groundwater contamination: Pump and treat remediation. Environmental Science and Technology 23(6):630-636.

Marley, M. C., and C. J. Bruell. 1995. In Situ Air Sparging: Evaluation of Petroleum Industry Sites and Considerations for Applicability, Design and Operation. American Petroleum Industry Publication No. 4609. Washington, D.C.

McCarty, P. L., and J. T. Wilson. 1992. Natural anaerobic treatment of a TCE plume, St. Joseph, Michigan NPL site. Pp. 47-50 in Bioremediation of Hazardous Wastes. EPA/600/R-92/126. Cincinnati, Oh.: EPA Center for Environmental Research Information.

McCarty, P. L, M. N. Goltz, G. D. Hopkins, M. E. Dolan, J. P. Allan, B. T. Kawakami, and T. J. Carrothers. 1998. Full-scale evaluation of in situ cometabolic degradation of trichloroethylene in groundwater through toluene injection. Environmental Science and Technology 32:88-100.

McCray, J. E., and R. W. Falta. 1996. Defining the air sparging radius of influence for groundwater remediation. Journal of Contaminant Hydrology. 24:25-52.

Mercer, J. W., and R. M. Cohen. 1990. A review of immiscible fluids in the subsurface. Journal of Contaminant Hydrology 6:107-163.

Miller, C. M., R. L. Valentine, M. E. Roehl, and P. J. Alvarez. 1996. Chemical and microbiological assessment of pendimethalin-contaminated soil after treatment with Fenton's reagent. Water Resources 30(11):2579-2586.

Montgomery, J. H. 1991. Groundwater Chemicals Field Guide. Chelsea, Mich.: Lewis Publishers.

Mueller, J. G., P. J. Chapman, and P. H. Pritchard. 1989. Creosote-contaminated sites. Environmental Science and Technology 23(10):1197-1201.

Newman, L. A., S. E. Strand, N. Choe, J. Duffy, G. Ekuan, M. Ruszaj, B. B. Shurleff, J. Wilmoth, P. Heilman, and M. P. Gordon. 1997. Uptake and biotransformation of trichloroethylene by hybrid poplars. Environmental Science and Technology 31(4):1062-1067.

Newmark, R. L., W. D. Daily, K. R. Kyle, and A. L. Ramirez. 1997. Monitoring DNAPL Pumping Using Integrated Geophysical Techniques. Report UCRL-ID-122215. Livermore, Calif.: Lawrence Livermore National Laboratory.

Norris, R. D., R. E. Hinchee, R. Brown, P. L. McCarty, L. Semprini, J. T. Wilson, O. H. Kampbell, M. Reinhard, E. J. Bouwer, R. C. Borden, T. M. Vogel, J. M. Thomas, and C. H. Ward. 1994. Handbook of Bioremediation. Boca Raton, Fla.: Lewis Publishers.

NRC (National Research Council). 1993. In Situ Bioremediation, When Does it Work? Washington, D.C.: National Academy Press.

NRC. 1994. Alternatives for Groundwater Cleanup. Washington, D.C.: National Academy Press.

NRC. 1996. Glass as a Waste Form and Vitrification Technology: Summary of an International Workshop. Washington, D.C.: National Academy Press.

Oma, K. H., D. J. Wilson, and A. N. Clarke. 1994. Pp. 457-492 in Hazardous Waste Soil Remediation: Theory and Application of Innovative Technologies. New York: Marcel Dekker.

Pankow, J. F., and J. A. Cherry. 1996. Dense Chlorinated Solvents and other DNAPLs in Groundwater. Portland: Waterloo Press.

Pardieck, D. L., E. J. Bouwer, and A. T. Stone. 1992. Hydrogen peroxide use to increase oxidant capacity for in situ bioremediation of contaminated soils and aquifers: A review. Journal of Contaminant Hydrology 9:221-242.

Peters, C. A., and R. G. Luthy. 1993. Coal tar dissolution in water-miscible solvents: Experimental evaluation. Environmental Science and Technology 27:2831-2343.

Pope, G. A., and W. H. Wade 1995. Lessons from enhanced oil recovery research for surfactant enhanced aquifer remediation. ACS Symposium Series 594:142-160.

Rao, P. S. C., A. G. Hornsby, D. P. Kilcrease, and P. Nkedi-Kizza. 1985. Sorption and transport of hydrophobic organic chemicals in aqueous and mixed solvent systems: Model development and preliminary evaluation. Journal of Environmental Quality 14(3):376-383.

Riley, R. G., J. M. Zachara, and F. J. Wobber. 1992. Chemical Contaminants on DOE Lands and Selection of Contaminant Mixtures for Subsurface Science Research. DOE/ER-0547T. Washington, D.C.: U.S. DOE Office of Energy Research.

Santamarina, J. C., and M. Fam. 1997. Dielectric permittivity of soils mixed with organic and inorganic fluids. Journal of Environmental and Engineering Geophysics 2.:37-51.

Schnarr, M. J., C. T. Truax, G. J. Farquhar, E. D. Hood, T. Gonullu, and B. Stickney. 1998. Experiments using potassium permanganate to remediate trichloroethylene and perchloroethylene DNAPLs. Journal of Contaminant Hydrology 29(3):205-224.

Schnoor, J. L. 1997. Phytoremediation. TE-98-01. Pittsburgh, Pa.: Groundwater Remediation Technologies Analysis Center.

Schwille, F. 1988. Dense Chlorinated Solvents in Porous and Fractured Media, Model Experiments. Chelsea, Mich.: Lewis Publishers.

Semprini, L. 1997a. In situ transformation of halogenated aliphatic compounds under anaerobic conditions. Pp. 429-450 in Subsurface Restoration, C. H. Ward, H. A. Cherry, and M. R. Scalf, Eds. Chelsea, Mich.: Ann Arbor Press.

Semprini, L. 1997b. Strategies for the aerobic co-metabolism of chlorinated solvents. Current Opinion in Biotechnology 8(3):296-308.

Shoemaker, S. H., J. F. Greiner, and R. W. Gillham. 1995. Permeable reactive barriers. In Assessment of Barrier Containment Technologies. R. R. Rumer and J. K. Mitchel eds. PB96-180583. Springfield, Va.: National Technical Information Service.

Siegrist, R. L., K. S. Lowe, L. D. Murdoch, W. W. Slack, and T. C. Houk. 1997. X-231 A Demonstration of In Situ Remediation of DNAPL Compounds in Low Permeability Media by Soil Fracturing with Thermally Enhanced Mass Recovery of Reactive Barrier Destruction. ORNL/TM-13534. Oak Ridge, Tenn.: Oak Ridge National Laboratory.

Steeples, D.W. 1998. Shallow Seismic Special Issue, Introduction. Geophysics 63:1210-1212.

Thornton, E. C., J. E. Szecody, K. J. Cantrell, C. J. Thompson, J. C. Evans, J. S. Fruchter, and A. V. Mitroshkov. 1991. Reductive dechlorination of TCE by dithionite-treated sediment. Pp. 335-340 in Remediation of Chlorinated and Recalcitrant Compounds: Physical, Chemical, and Thermal Technologies, G. B. Wickramanayake and R. E. Hinchees, Eds. Columbus, Oh.: Battelle Press.

Udell, K. S. 1997. Thermally enhanced removal of liquid hydrocarbon contaminants from soils and groundwater. Chapter 16 in Subsurface Restoration, C. H. Ward, J. A. Cherry, and M. R. Scalf, Eds. Ann Arbor, Mich.: Ann Arbor Press.

U.S. Air Force. 1996. Technology Performance and Application Analysis of In Situ Air sparging, Operable Unit 6, Hill Air Force Base, Utah, 19 pp.

Vidic, R. D., and F. G. Pohland. 1996. Treatment Walls. Ground-Water Remediation Technologies Analysis Center, Technology Evaluation Report. TE-96-01. Pittsburgh, Pa.: Ground-Water Remediation Technologies Analysis Center.

Vogel, T. M., C. S. Criddle and P. L. McCarty. 1987. Transformation of halogenated aliphatic compounds. Environmental Science and Technology 21:722-736.

Wiedemeier, T. H., M. A. Swanson, D. E. Moutoux, E. K. Gordon, J. T. Wilson, B. H. Wilson, J. H. Kampbell, J. E. Hansen, P. Haas, and F. H. Chapelle. 1997. Technical Protocol for Evaluating Natural Attenuation of Chlorinated Solvents in Groundwater. San Antonio, Tex.: Air Force Center for Environmental Excellence, Brooks Air Force Base.

Wilson, D. J., and A. N. Clarke. 1994. Hazardous Waste Soil Remediation: Theory and Application of Innovative Technologies. New York: Marcel Dekker, Inc.

Wilson, J. T., J. W. Weaver, and K. D. H. Kampbell. 1994. Intrinsic bioremediation of TCE in groundwater at an NPL site in St. Joseph, Michigan. In Symposium on Intrinsic Bioremediation of Groundwater. EPA/540/R-94/515. Washington, D.C.: U.S. Environmental Protection Agency, Office of Research and Development.

Wilson, J. T., and B. H. Wilson. 1985. Biotransformation of trichloroethylene in soil. Applied and Environmental Microbiology 49:242-243.

Young R. A., and J. Sun. 1996. 3D ground penetrating radar imaging of a shallow aquifer at Hill AFB, Utah. Journal of Environmental and Engineering Geophysics 1:97-109.

5

DOE Remediation Technology Development: Past Experience and Future Directions

When the Department of Energy (DOE) established the Subsurface Contaminants Focus Area (SCFA) in the mid-1990s, few innovative technologies were used to clean up contaminated groundwater and soil at DOE installations. For example, as of 1995, the only innovative remedy specified for groundwater cleanup at DOE sites regulated under the Comprehensive Environmental Response, Compensation, and Liability Act (CERCLA) was one application of natural attenuation,[1] according to Environmental Protection Agency (EPA) data. Only soil vapor extraction (SVE) had significant application for contaminated soil (see Table 5-1).

This chapter assesses SCFA's recent progress in developing and deploying new technologies for cleaning up contaminated groundwater and soil. The chapter first reviews barriers in transferring SCFA technologies from the research and development stage to full-scale deployment. The chapter then reviews the extent to which innovative methods have been applied in the cleanup of groundwater and soil at DOE installations and the extent to which groundwater and soil remediation technologies developed by SCFA have been used. The chapter concludes with a review of steps that SCFA has taken to improve its process for selecting which technologies to develop. Also included are descriptions of several recent successful SCFA technology development projects, which can provide models for planning future projects.

---

[1] At the time of this study, 1995 was the most recent year for which data were available on technologies specified in CERCLA records of decision.

TABLE 5-1 Use of Innovative Technologies at DOE Sites Regulated Under CERCLA

Contaminated Medium	Total Number of Sites	Number with Conventional Remedy	Other Remedies
Groundwater	13	11	1 natural attenuation
			1 institutional controls only
Soil	17	12	4 soil vapor extractions
			1 excavation with ex situ solidification or stabilization
			1 cover with clean soil

SOURCE: EPA, 1997.

BARRIERS TO INNOVATIVE TECHNOLOGY USE AT DOE SITES

The DOE's Office of Science and Technology (OST), under which the SCFA operates, has been criticized for failing to organize a research program that leads to significant applications of innovative remediation technologies. However, DOE is not alone in its limited application of innovative remediation technologies. In the cleanup of contaminated groundwater and soil at privately owned CERCLA sites, for example, application of innovative technologies historically has been limited. According to EPA data, innovative remedies had been selected for contaminated groundwater at only 6 percent of all CERCLA sites as of 1995 (EPA, 1996). Innovative technologies other than SVE had been selected for only 26 percent of all soil cleanup under CERCLA (EPA, 1996). DOE's historical problems in deploying innovative remediation technologies thus have parallels in other sectors.

Lack of Demand

A recent National Research Council (NRC, 1997a) study of innovative remediation technologies in the private sector concluded that lack of customer demand was the primary obstacle to more rapid technology development. The NRC attributed this lack of demand to insufficient incentives for the prompt cleanup of contaminated sites. The NRC report concluded, "A major failing of national policy in creating a healthy market for environmental remediation technologies is the lack of sufficient mechanisms linking the prompt cleanup of contaminated sites with the financial self interest of the organization responsible for the contamination." As a result of this lack of demand, the NRC found, small remediation technology development companies have struggled to stay in business. For example, the stock value of the seven private-sector remediation technology companies that have gone public has decreased, in most cases precipitously, since the initial public offering (MacDonald, 1997).

OST, and within it SCFA, is analogous to a small technology development firm within DOE and has fared similarly to its private-sector counterparts. Customer demand for SCFA's technologies is lagging in part because of a historical lack of financial incentives for the rapid cleanup of contaminated DOE facilities. On the contrary, rapid cleanup of DOE sites can lead to loss of revenue for the DOE site management contractor and loss of local jobs once the cleanup is completed and the site closed (GAO, 1994a). Contractors and managers at DOE installations have resisted efforts by DOE headquarters and OST to "push" the use of innovative technologies.

In a 1995 review of federal agency efforts to clean up contaminated sites, the U.S. General Accounting Office (GAO) concluded that inadequate contract management was a major reason for the slow progress in site cleanups (Guerrero, 1995). Slow progress in cleanup, in turn, limits demand for innovative remediation technologies. GAO concluded, "DOE's problems were compounded by its failure to ensure the effective oversight of its contractors' financial management." Site management contractors could be fully reimbursed for charges incurred in site cleanup activities, but DOE's oversight of these charges was inadequate, according to GAO. One study concluded that poor contract management had increased DOE's cleanup costs 32 percent above those in the private sector and 15 percent above those in other federal agencies (Guerrero, 1997).

DOE data confirm that a major barrier to the use of innovative remediation technologies is the failure of site managers to seek applicable innovative technologies. Table 5-2 shows the results of a survey of 232 DOE sites where innovative remediation technologies were not selected for application. At 71 of these sites, project managers automatically chose the baseline without identifying innovations. At 85 sites, they indicated that no applicable innovative technologies were available, which also might be attributed to failure to search for alternative technologies.

Many demonstrations of innovative remediation technologies have occurred at DOE sites, but in the past these demonstrations were seldom converted to full-scale cleanup operations. According to SCFA managers, DOE site management contractors received significant funding for conducting innovative technology demonstrations, which created an incentive to field test numerous technologies in order to bring additional rev-

TABLE 5-2 Reasons for Not Selecting Innovative Technologies for Remediation of Contaminated DOE Sites

Reason	Number of Sites
Not cost-effective	68
Baseline technology selected; no innovations identified	71
No applicable innovative technologies	85
Innovative technology has become the baseline	6
Perceived regulatory resistance to innovations	2
Total	232

SOURCE: Data submitted by DOE's OST in response to questions from Representative Bliley, September 24, 1997.

enue to the site. However, the lack of sufficient incentives to complete cleanups, plus the risk that the contractor might incur the additional liability of constructing a conventional cleanup system if the innovative one failed at full scale, provided major disincentives to full-scale deployment (GAO, 1994a).

Much of the reason for the lack of innovative remediation technology at DOE sites is thus external to SCFA management. Lack of demand for innovative remediation technologies from individual field sites is a major barrier to the application of innovative technology. This problem is not unique to DOE and has parallels in the private sector.

Other Barriers

Other barriers to innovative remediation technology development and application also exist within DOE, and OST and SCFA have taken steps to address some of these. The other barriers can be grouped into four categories: (1) shortcomings in OST planning and management, (2) insufficient involvement of technology end users in setting technology development priorities, (3) public resistance to innovative technology use, and (4) regulatory requirements that favor conventional technologies.

Reports by the GAO (1992, 1994a, 1996a) have identified OST management problems as one reason for the slow development of innovative remediation technologies within DOE. The 1992 report concluded that OST lacked clear decision points for deciding when to continue funding research projects and when to terminate them. Also lacking at that time were cost estimates, project development schedules, and measurable performance goals for research projects receiving OST funding. As a result of these deficiencies, GAO concluded, OST lacked mechanisms for eliminating poorly performing projects and measuring overall program performance. Flaws with OST management identified in the 1994 GAO report included lack of a comprehensive technology needs assessment to guide research priorities and lack of a coordinated mechanism for identifying available technical solutions. The GAO found that other offices within DOE's environmental management program were funding technology research that overlapped with OST's. The 1996 GAO report pointed to lack of coordination among DOE remediation technology development activities, plus favoritism in selecting research projects for certain sites, as remaining problems with OST management. As discussed later in this chapter, OST has recognized these problems and is responding accordingly.

Insufficient involvement of end users (the customers for innovative technologies) in SCFA's technology development program is another important factor that has curtailed deployment of innovative remediation technologies developed by SCFA. In a 1998 review of the extent to which innovative technologies developed by OST have been deployed, GAO concluded that lack of end user involvement is one of the major remaining obstacles to more widespread use of technologies developed by OST as a whole (GAO, 1998b). GAO concluded that OST has not sufficiently involved the DOE field site personnel responsible for restoration activities in the technology development decision-making process. In addition, OST has not provided for sufficient involvement of field site personnel in individual technology development projects.

Site regulators and vocal members of the public have also limited the application of innovative remediation technologies at DOE sites, according to some reports (GAO, 1994a; Nemeth et al., 1997). Local officials and regulators may fear that an innovative technology has a less certain chance of meeting cleanup milestones than a conventional one (GAO, 1994a) and therefore may deny approval to use the innovative technology. Members of the public near contaminated sites may oppose use of innovative technologies for similar reasons. Regulators may hesitate to appear lenient before an active public by allowing the use of a less costly technology whose performance is uncertain.

The regulatory requirements for selection of cleanup remedies under CERCLA and the Resource Conservation and Recovery Act (RCRA) also have been faulted for limited use of innovative technologies. CERCLA requires consideration of nine evaluation criteria (listed in Box 2-2) when selecting the final remedy for a site, and the RCRA remedy selection process generally parallels CERCLA. The first two criteria, which require that the selected remedy be protective of human health and meet applicable requirements of other regulations, are the critical ones that regulators consider and do not necessarily favor conventional remedies. However, the remaining seven criteria require evaluation of a record of cost and performance data for the technology. These criteria create a bias toward conventional cleanup technologies, because judging whether a technology will meet the criteria requires a preexisting record of performance. For many innovative technologies, cost and performance data for large-scale implementation are lacking, making it difficult to judge whether these technologies meet the criteria.

DOE STEPS TO INCREASE INNOVATIVE TECHNOLOGY DEPLOYMENT

DOE managers are now well aware of many of the impediments to remediation technology development and have taken steps to address these problems. OST instituted a variety of management reforms (including efforts to involve end users in its decision process) in response to criticism from the GAO, for example. In addition, the DOE Office of Environmental Management and OST have worked to decrease regulatory resistance to using innovative remediation technologies. More recently, DOE began implementing a new contracting approach for contaminated site cleanups that, in theory, includes incentives for completing cleanup on time and on or under budget.

Among the most important OST management reforms is a change in the process used to decide which technology development projects should receive funding (NRC, in review). During OST's inaugural years, in the early 1990s, funding decisions were made by the head of OST with essentially no involvement of those who would ultimately be the "customers" for the technologies that OST was developing. By 1994, however, OST recognized the need to shift to a decision process that would include formal involvement of technology end users.

To provide a mechanism for involving technology end users in its funding decision process, OST established a team for each major installation to identify the installation's primary needs for completing cleanup work. These teams, known as site technology coordination groups (STCGs), consist of personnel from the installation's DOE operations office, operating contractor's office, and laboratories. Under OST's current funding decision process, STCGs submit statements describing their needs to the appropriate office within OST (such as SCFA). OST then groups the needs into like categories and further groups the categories into "work packages." Table 5-3 shows SCFA's 1999 work package list; this list was developed by consolidating the STCG needs statements. OST next solicits proposals to fill the technology gaps as identified in the work packages. To determine which proposals will be funded, OST managers work with the STCGs and other interested stakeholders (such as regulators) to develop criteria for determining funding priorities within each work package. Figure 5-1 shows the priority-setting matrix used in 1998; the num-

TABLE 5-3 SCFA Work Packages for 1999

Package Number	Package Title
1a	DNAPL contamination
1b	Access-in situ metals-radionuclides treatment
2	Containment-stabilization
3	Delineation: complex or deep
4	Treatment delivery systems
5	Source-term remediation
6	Containment structures (>30 m [100 ft])
7	Metals-rad mobilization-extraction technologies
8	Tritium containment
9	Delineation geophysics (15-30 m [50 to 100 ft])
10	Explosive-pyrophoric materials

NOTE: Work packages are listed in priority order, DNAPL = dense nonaqueous-phase liquid.

SOURCE: Baum, 1998a.

bers in each box indicate the relative weight given to each criterion listed at the left of the matrix.

Other OST management changes include the following:

- *Implementation of a "gate" process for project decision making* (see Figure 5-2). OST established a gate review system to address the problem of lack of clear decision points for determining when to continue or terminate project funding. The six gates, as shown in Figure 5-2, represent points at which funding and other decisions are made. They are based on the investment decision model presented in *Winning at New Products* (Cooper, 1993). The model depicts technology development as encompassing seven stages, from basic research (stage 1) through commercialization (stage 7). OST's six "gates" represent the passage from one of Cooper's stages to the next.
- *Tracking of cost estimates and deployment schedules for each project.* OST established an automated central tracking system with information on schedules and costs for OST-funded projects. This system was designed in response to a GAO report indicating that OST lacked basic management tools, including a tracking system (Rezendes, 1997).
- *Preparation of a comprehensive list of remediation technology development projects within DOE.* OST developed a list to identify overlapping efforts that could be cut or combined to reduce duplication. For example, a GAO (1996a) review of OST indicated that in 1996, DOE was fully funding studies of vitrification systems at 52 sites across the country, with little

SCFA Priority-Setting Matrix

Work Package Title:________________________________Date:____________
Assumptions:____

Criterion Number	Criterion	Very High (4)	High (3)	Medium (2)	Low (1)	Fail (0)
Ten-Year Plan Applicability (60 pts possible)						
1	Prevalence in DOE Complex	15	12	7	4	0
2	DOE End User Commitment	15	12	7	4	0
3	Cost Reduction	15		7	4	0
4	Ability to Meet Compliance	15		7	4	0
Performance (40 pts)						
5	Technical Viability	5		3		0
6	Engineering/ Constructibility	5		3		0
7	Business Performance	15	12	7	4	0
8	Risk Reduction (public, worker, environment)	5	4		1	0
9	Stakeholder Acceptance	10		5		0
Column Totals						

Total:_______ Scorers' Names:____________________________________

FIGURE 5-1 SCFA matrix used to rank proposals submitted for funding.

coordination among the projects; a year later, as part of the effort to reduce duplication, OST cut the number of such studies to five (GAO, 1996a; Rezendes, 1997).

- *Institution of an independent peer review process.* OST has instituted a peer review process, overseen by the American Society of Mechanical Engineers, to provide independent evaluations of select technology development projects. However, this program is still evolving, and peer review is not yet an integral part of every technology development project (NRC, 1997b).

OST has also instituted programs for decreasing regulatory resistance to the use of innovative remediation technologies. Working with the Southern States Energy Board, OST has organized a series of technology

Technology Maturation Stages	Basic Research	Applied Research	Exploratory Development	Advanced Development	Engineering Development	Demonstration	Implementation
Activities	Idea Generation		Proof of Technology		Engineering Prototype	Production Prototype	Utilization by end user
	Need	Need	Product definition Nonspecific applications Bench scale	Working model Reduction to practice Specific applications Bench scale	Scaled-up versions to test design feature and performance limits Pilot scale Field testing	End user validation Full scale "Beta" site testing	
Gates	**1**		**2**	**3**	**4**	**5**	**6**
Expectations		Address priority DOE need Knowledge of similar efforts		Show clear advantage over available technology	Meet cost-benefit requirement Demonstrate significant end user demand	Technology ready for end user	End user deploys technology

FIGURE 5-2 Six gates used as decision points for continuing or discontinuing project funding under OST. SOURCE: Hill et al., 1997.

demonstrations in which regulators are directly involved in the planning (Nemeth et al., 1997). For each such demonstration, a team is appointed to establish remediation goals and define the market for the technology. The team consists of federal regulators, state regulators, technology developers, representatives of DOE sites, financiers, members of the public, a U.S. Army Corps of Engineers innovative technology advocate, and representatives of the Southern States Energy Board and the Western Governors' Association. At the end of the demonstration, the involved state and federal regulators sign a statement verifying the technology's performance, if it was successful. The verification statement can then be used to reduce future regulatory approval requirements or to satisfy potential users that the technology will perform as advertised.

In addition to these efforts by OST, DOE has undertaken contracting reforms and developed financial incentives designed in part to accelerate cleanup of contaminated sites. Providing incentives for rapid cleanup would, in turn, increase demand for new cost-effective remediation technologies. Beginning in 1994, DOE instituted the "Contract Reform Initiative" to address inefficiencies resulting from the department's historical contracting practices. Historically, a single contractor at each DOE installation carried out most environmental cleanup and other operations under a cost-reimbursible contract in which the contractor not only was paid for the expenses of running the installation but also was awarded a profit. This type of contracting arrangement not only lacked specific incentives for completing major site cleanup tasks but also created hidden disincentives for completing the cleanups because contractors would lose their jobs once the cleanup was complete. Under the Contract Reform Initiative, DOE has developed a new type of contracting procedure known as the performance-based management contract. This type of contract ties the contractor's profit to achieving specific milestones related to DOE's overall goals for completing site cleanup. Under the reform initiative, DOE is also increasing the use of competitive bidding in awarding contracts. In addition, at some installations, DOE is using an approach known as "management and integration" contracting, in which cleanup work is performed by a team of subcontractors overseen by a prime contractor. Another important component of the new contracting approach is the increasing use of fixed-price contracts. These and other reform measures are designed to create market pressure to complete site cleanup.

In fiscal year 1998, the Office of Environmental Management also established a new financial awards program to create incentives for using innovative remediation technologies. The program, known as the Accelerated Site Technology Deployment Program, provides funds for the first site that uses an innovative technology. The program is not designed to support demonstrations of new technologies but rather to support first-time, full-scale applications of technologies that have undergone sufficient pilot testing to generate cost and performance data (NRC, in review). Under the program, managers of individual DOE site cleanup projects can apply for funds for first-time use of an innovative technology provided they can show the level of cost savings expected in comparison to application of the baseline technology. Funding for this initiative was $25 million for fiscal year 1998 (NRC, in review). Table 5-4 lists projects funded under the Accelerated Site Technology Deployment Program in 1998.

DEPLOYMENT OF INNOVATIVE REMEDIATION TECHNOLOGIES AT DOE INSTALLATIONS

According to data from SCFA, 146 deployments of 56 innovative technologies developed by SCFA had occurred as of January 14, 1998 (see Appendix B). This large number appears to be a dramatic improvement since 1995. However, whether this signifies a major step forward in deploying SCFA-tested and -developed innovative technologies is uncertain, primarily for four reasons.

First, site data from DOE's Office of Environmental Restoration do not confirm that a large number of innovative technologies are being used for full-scale cleanup of groundwater and soil at DOE installations. As indicated in Tables 5-5 and 5-6, the range of technologies being used in actual cleanup projects at DOE installations as reported by DOE remediation project managers in the summer of 1997 is quite limited and does not include many of SCFA's innovations. For example, the predominant remedies for groundwater as reported by project managers are pump-and-treat systems (used in 41 percent of the projects), natural attenuation (used in 22 percent), and capping and containment (used in 19 percent). These data do not reflect the use of innovative site characterization technologies, because site characterization technology use is not reported to the Offfice of Environmental Restoration. The data are also about a year less current than the SCFA deployment list. Nonetheless, the data appear to indicate that the range of technologies being used for groundwater and soil cleanup is still relatively limited. It is doubtful that there has been a surge in use of innovative remediation technologies since these data were compiled, given the long period required for remedy selection at most sites.

Second, SCFA's list of innovative technology deployments to date indicates a lack of multisite applications for most technologies (see Table 5-7). Although 29 (52 percent) of the 56 technologies have been deployed at more than one facility, only 10 (18 percent) have been deployed at more than two facilities. Ideally, to save money and advance cleanup progress,

TABLE 5-4 Projects Funded Under the Accelerated Site Technology Deployment Program in Fiscal Year 1998

Project Title and Location	Description
Alternative landfill cover system; deployment of the mixed waste landfill, Sandia National Laboratories	An evapotranspiration cover, in lieu of a more expensive RCRA cover, combined with an innovative fiber-optic monitoring system that eliminates the need for long-term groundwater monitorng will be deployed on an actual radioactive mixed-waste landfill.
Decontamination and volume reduction system, Albuquerque	A combination of technologies will be used to decontaminate and reduce the volume of transuranic-contaminated gloveboxes in storage.
Deployment of a permeable reactive treatment wall for radionuclides and metals, Albuquerque	This treatment wall will contain a barrier impermeable to certain radioactive materials and a permeable "gate" that will allow other materials to pass through.
In-well air stripping technology to remediate an off-site organic plume, Brookhaven National Laboratory	In-well air stripping will remove VOCs from groundwater in situ through mass transfer to the air phase in the well.
Development of an integrated technology suite for cost-effectively delineating contamination in support of soil remediation actions, Fernald Environmental Management Project	Technologies will improve detection and location of gamma-emitting radionuclides in surface soils and provide data analysis for decision support.
Integrated decontamination and decommissioning (combined with) decontamination and decommissioning of 29 structures at Idaho and Fernald	A suite of technologies that have been successfully demonstrated at previous D&D large-scale demonstration projects will be deployed.
Contaminated soil cleanup using the ACT*DE*CON[SM] process, Ohio (Mound)	The ACT*DE*CON process will be used for cleanup of soils or sediments contaminated with radionuclides and heavy metals that must be transported and disposed .

continues on next page

TABLE 5-4 Continued

Project Title and Location	Description
Segmented gate system (combined with) integrated soil processing, Ohio (Mound) and Idaho	The segmented gate system will mechanically separate radioactively contaminated soil into a clean stream and a contaminated stream.
Improved systems for tank sludge retrieval, conditioning, and transfer, Oak Ridge	Technologies developed by private industry participants will be used to enhance mixing and mobilization of sludge from various tank configurations and subsequently transfer it to intermediate storage tanks and/or treatment facilities.
Electrochemical ion exchange for waste reduction (combined with) modular evaporator system for waste volume reduction in tanks, Oak Ridge	A modular single-stage subatmospheric system will concentrate liquid radioactive waste prior to immobilization for disposal. Also, a proposed highly selective crystalline silicotitanate ionexchange process will remove radiological contaminants.
Slurry monitoring, Richland	New instruments will reduce the probability of pipeline blockage due to solids segregation, crystallization, and gelation, and therefore unnecessary line replacement costs.
Enhanced in situ decontamination and size reduction of gloveboxes, Rocky Flats	Combination of in situ technologies that provides for the radiological characterization, decontamination, and site reduction of transuranic-contaminated materials.
Fluidic sampler, Savannah River	Two fluidic samplers will be deployed to obtain accurate and representative samples from sludge feed tanks containing high-level waste.

NOTE: VOC = volatile organic compound.

SOURCE: Accelerated Site Technology Deployment program, http://wastenot.inel.gov/tdi.

TABLE 5-5 Technologies Used to Clean up Groundwater at DOE Projects

Technology	Number of Projects
Pump and treat	11
Natural attenuation or intrinsic bioremediation	6
None	3
Cap	3
Containment stystem	2
Air sparging	2
Free product recovery	2
Thermally enhanced vapor extraction	1
Passive reactive barriers	1
No data available	1

NOTE: The total number of projects represented by these data is 27, but some projects involve more than one technology.

SOURCE: M. Tolbert-Smith, U.S. Department of Energy, Office of Environmental Restoration, July 16, 1998 (based on data as reported by DOE remediation project managers).

TABLE 5-6 Technologies Used to Clean up Soil at DOE Projects

Technology	Number of Projects
Excavation, followed by disposal, ex situ treatment, or storage	98
Solidification or stabilization with cement or grout	32
Passive treatment wetlands	10
Caps	9
Natural attenuation	8
Land farming or ex situ bioremediation	4
Soil vapor extraction or bioventing	4
Thermally enhanced vapor extraction	1

NOTE: The total number of projects represented by these data is 163, but some projects involve more than one technology.

SOURCE: M. Tolbert-Smith, U.S. Department of Energy, Office of Environmental Restoration, July 16, 1998 (based on data as reported by DOE remediation project managers).

TABLE 5-7 SCFA Technologies Deployed at More Than One DOE Facility

Technology	Number of Facilities Where Deployed	Total Number of Deployments at DOE Sites
Six-phase soil heating	2	2
In-well vapor stripping (recirculating wells)	3	4
Dig-face characterization	2	3
Passive reactive barrier	2	2
Thermal enhanced vapor extraction system	3	4
SEAMIST	4	5
Deep-soil mixing	2	3
Resonant sonic drilling	4	7
Passive SVE	3	4
In situ permeable flow sensor	2	2
Cryogenic cutting	2	2
In situ anaerobic bioremediation	2	3
In situ chemical oxidation (soils)	2	4
Adsorption or desorption relative to in situ bioremediation of chlorinated solvents	2	2
Colloidal borescope	2	3
Fiber-optic probe for TCE in groundwater	2	3
Heavyweight cone penetrometer	3	10
Rapid transuranic monitoring laboratory	2	3
Advanced in situ moisture logging	2	3
Field screening laboratory system	2	2
Remote excavation system	2	2
Absorptive stripping voltametry	2	3
Cross-well seismic imaging	2	4
Long-range alpha detector	3	4
Cross-hole compressional and shear wave seismic tomography	2	2
Directional drilling	5	8
Electromagnetic geophysical surveyor	3	4
Micropurging of wells	4	5
Rapid geophysical surveyor	2	2
Number deployed at > 1 facility	29	
Number deployed at > 2 facilities	10	

NOTE: As explained in Chapter 1, "facility" refers to an entire installation, such as Hanford or Los Alamos. "Site" refers to an individual contaminated area, such as a plume of contaminants in groundwater within a facility. One facility may contain many contaminated sites. TCE = trichloroethylene.

SOURCE: Data provided by SCFA (see Appendix B).

considerable technology development work should be directed at systems that can be adopted across the weapons complex.[2]

Third, fewer than one-third (18 of 56) of the technologies on SCFA's deployment list address the most critical need related to subsurface cleanup: in situ remediation of contaminants in groundwater and soil. The remaining technologies are for site characterization and monitoring. DOE managers should evaluate whether the development of in situ remediation technologies is being given appropriate priority.

Fourth, many of the listed remediation technologies were developed outside DOE. Technology development occurs in a variety of institutions. For example, passive reactive barriers were developed by researchers at the University of Waterloo; in situ bioremediation was developed largely by the petroleum industry in 1972; and in situ chemical oxidation was developed by private companies that hold patents on this technology (Brown et al., 1993; NRC, 1997a). Many technology projects, such as development of the Lasagna® process by a government-industry consortium, are collaborative initiatives. Work on many of the technologies on the deployment list in Appendix B (including reactive barriers, in situ bioremediation, and in situ chemical oxidation) occurred in the private sector or in other agencies, as well as in SCFA. Although SCFA should be commended for pursuing collaborative technology development projects and for adapting technologies to DOE problems because these activities can leverage limited financial resources, determining SCFA's role in furthering the development of these systems is difficult. Further, the inclusion of these technologies within the SCFA deployment list suggest a tendency for SCFA to "reinvent" existing technologies rather than support existing innovators. The GAO reached a similar conclusion and noted that "OST staff are not always well-informed about technologies developed by organizations other than OST" (GAO, 1998b). This tendency to "reinvent the wheel" results in a significant amount of research within DOE that closely parallels previous external research. Replication of external research results in the inefficient use of resources and potentially in infringement of intellectual property rights, creating a lack of good will between DOE and technology developers. In addition, it demonstrates a lack of sufficient effort within DOE to enlist the participation of leaders in the field of remediation technology development.

The data provided by SCFA are thus not yet sufficient to assess whether OST's management reforms have led to SCFA technologies hav-

---

[2] A technology's design specifications and applicability will vary by site, and some problems unique to a given facility may be sufficiently critical in terms of risk and cost to warrant the development of technology without multisite application.

ing a greater impact on the cleanup of DOE facilities. The GAO found similar problems with OST innovative technology deployment data in its 1998 review. According to GAO's report, "GAO found many errors in the office's [OST's] deployment data. . . . The Office of Science and Technology overstated its deployment information."

EFFECTIVENESS OF REFORMS IN PROMOTING DEPLOYMENTS

What effect the OST and Office of Environmental Management reforms and initiatives will have on innovative remediation technology deployment at DOE sites in the near future is uncertain. The steep cuts in SCFA's budget present a critical obstacle to promoting deployment of SCFA technologies. As described in Chapter 1, the SCFA budget has decreased from a high of $82.1 million in 1994 to approximately $10 million in fiscal year 1998 (after discounting congressional earmarks). A $10 million budget is insufficient to support the types of large-scale field demonstrations necessary to advance the use of innovative technologies. Further, because of these budget cuts, the SCFA program has been fully mortgaged since 1996 (Baum, 1998a,b), meaning that the full budget is used to support multiyear projects that were slated for funding before SCFA was formed. In 1998, SCFA carried out the formal process of soliciting needs statements from the STCGs, categorizing these into work packages, and prioritizing projects, but it was able to apply this process only to existing projects. SCFA solicited proposals from the national laboratory personnel already funded under the program as if they were competing for reentry into the program (Baum, 1998a,b). Whether SCFA can succeed in implementing program reforms and increasing its influence on the effectiveness of the DOE cleanup program with its current budget is unclear.

Although SCFA's budget has been cut, some have looked to the Accelerated Site Technology Deployment Program to increase the deployment of SCFA-developed innovative technologies. However, this program also might not have a sufficient budget to succeed. The amount of money provided ($25 million), although greater than the SCFA budget, is approximately equal to the average cost of cleaning up one CERCLA site. These funds will be divided among many sites. The amount each site receives may not be significant enough to encourage contractors to risk deploying an innovative technology unless performance of the innovative technology is guaranteed (NRC, 1997a). Further, whether the technology will be deployed a second time, given the fact that only the first user receives funding for the deployment, is uncertain (Rezendes, 1997).

Whether DOE's broad environmental contracting reforms will succeed and will increase the likelihood of SCFA technology deployment is also unclear. A 1997 assessment of DOE's contract reform measures by the DOE Office of the Inspector General identified major problems with the reform effort, including failure to link DOE's overall cleanup goals to specific financial incentives being granted under the new contracts and lack of guidance from headquarters on appropriate fee structures for different types of incentives (DOE, 1997a). A more recent review by the GAO indicated that DOE is working to address these problems but that it is too early to assess the overall effectiveness of the contract reform efforts (GAO, 1998a).

Further, under the fixed-price contracting approach being implemented as one part of the reform effort, DOE officials are removed from the technology selection process, making the link between OST and site decision makers even more tenuous. In the past, GAO has identified the lack of involvement of OST's technology developers in site cleanup technology decision making as a shortcoming (GAO, 1994a). In contrast, industry in general has found that a successful approach in terms of risk reduction and cost control is to provide a central organization, which includes technology developers, with the major role of establishing technologies and expenditures for remediation at all sites. This strategy allows risks to be prioritized among all sites, the highest-risk sites to be cleaned first, and the most cost-efficient technology to be applied. SCFA's efforts to involve technology end users in its program also have achieved limited success.

Although end users are now theoretically involved in setting SCFA programs direction through the STCG's, according to GAO the actual influence of these end users in OST's program as a whole has been quite limited. GAO concluded that a "rigorous application" of requirements to involve end users at various points in deciding whether to fund specific technology projects "might indicate that some projects should be terminated for reasons such as the lack of an identified customer" (GAO, 1998b). Further, according to GAO, end users still are not sufficiently involved in planning individual technology development projects, and as a result, end users report that OST (and SCFA) technologies are too generic to meet the needs of individual sites.

In summary, OST and SCFA have taken important steps to reform their programs, but the degree to which SCFA technologies are being deployed at full scale, and whether the reforms will succeed in increasing deployments, are unclear. More rapid progress in transferring SCFA technologies to full-scale field operations depends in large part on improving site contracting mechanisms in the DOE environmental restoration program as a whole, creating incentives for using innovative technologies, improving remediation technology, improving decision-making procedures, and providing for greater involvement of technology end users in setting SCFA's program direction. SCFA's future progress depends, as well, on the adequacy of its budget.

A recent cost savings review by the U.S. Army Corps of Engineers (1997) concluded that substantial cost savings, approximating $20 billion, can be realized from OST technologies, including 12 that were developed or enhanced with SCFA funding.[3] The Corps of Engineers concluded that standardized cost and performance reports are needed and that savings can be realized only through aggressive deployment of technologies. Clearly, the recognition of and demand for SCFA technologies to address problems that currently cannot be solved or cannot be solved in reasonable time frames or at reasonable cost have to be increased.

SCFA TECHNOLOGY DEVELOPMENT ACHIEVEMENTS

Despite the difficulties SCFA has faced in transferring its technologies to full-scale field operations, some SCFA technologies have shown considerable promise. This section highlights several successful SCFA projects to develop technologies for cleanup of metals, radionuclides, and DNAPLs in groundwater and soil. The committee used the technology reviews in Chapters 3 and 4 and information presented by SCFA technology developers at committee meetings to identify projects that have resulted in successful field demonstrations and full-scale applications. The committee did not analyze in detail all of SCFA's past technology development projects or attempt to rate them on a precise scale of success. Rather, these examples are intended to provide models for SCFA to follow in its future work and to show that this program has, in fact, led to some positive results.

Various metrics can be considered when attempting to assess the relative success of a particular technology or combination of technologies. Indices of success can range from the advancement of science and technology, which represents success at the level of proving fundamental principles, to timeliness and cost-effectiveness in reaching desired cleanup end points, which represents success in the application of demonstrated principles. As indicated in Box 5-1 (see also Figure 5-3), it is possible to identify the essential features of successful projects in order to provide

---

[3] The 12 SCFA technologies included in the Corps of Engineers review were (1) dynamic underground stripping; (2) passive soil vapor extraction; (3) barrier technologies (viscous liquids, frozen soil, and horizontal subsurface); (4) hybrid directional boring and horizontal logging; (5) in-well vapor stripping; (6) in situ bioremediation; (7) Lasagna®; (8) recirculating wells; (9) thermal enhanced vapor extraction; (10) in situ vitrification; (11) automated waste handling; and (12) landfill containment.

BOX 5-1
Learning From Successful Projects

The dig-face characterization project led to the development of a multisensor apparatus to allow real-time monitoring to determine the extent of contamination in a site being excavated, thereby guiding the program of excavation. The characterization system consists of on-site hardware for collecting detailed information on the changing chemical, radiological, and physical conditions in the subsurface soil during the entire course of a hazardous site excavation (see Josten et al., 1995).

The essential features of the project that made it a success and that can be broadly applied include the following:

- early, wide input and peer evaluation;
- clear advantage over current practice;
- multiple customers with high interest;
- significant benefits in cost savings, effectiveness, and safety;
- easy deployment and operation;
- reliability and robustness;
- quick adaptation to changes in unique and site-specific needs;
- enlistment of key expertise as project needs evolve;
- frequent and effective communication between principal investigators and end users;
- publication of technical results; and
- protection of intellectual property.

guidance for the planning, conduct, and assessment of other projects. In assessing SCFA projects, the factors the committee considered most important were whether (1) the project responded to a recognized and well-defined contamination problem identified by DOE field personnel; (2) initiation of the project was timely in responding to this problem; (3) laboratory and pilot-scale assessments were conducted to refine the technology; (4) the project resulted in well-defined design and operation parameters for the technology; (5) the technology resulted in cost savings and has the potential for multiple applications; (6) the project was independently peer reviewed; and (7) the project or technology met or is likely to meet the concerns of DOE site remediation managers, environmental regulators, and concerned members of the public. The committee also considered the availability and uniqueness of the technology, its stage of deployment, and the degree of interagency collaboration in technology development. The committee did not devise objective scales for evaluating technologies according to these criteria. Rather, it evaluated SCFA projects subjectively with these criteria in mind and selected by consensus examples that satisfy several of the criteria.

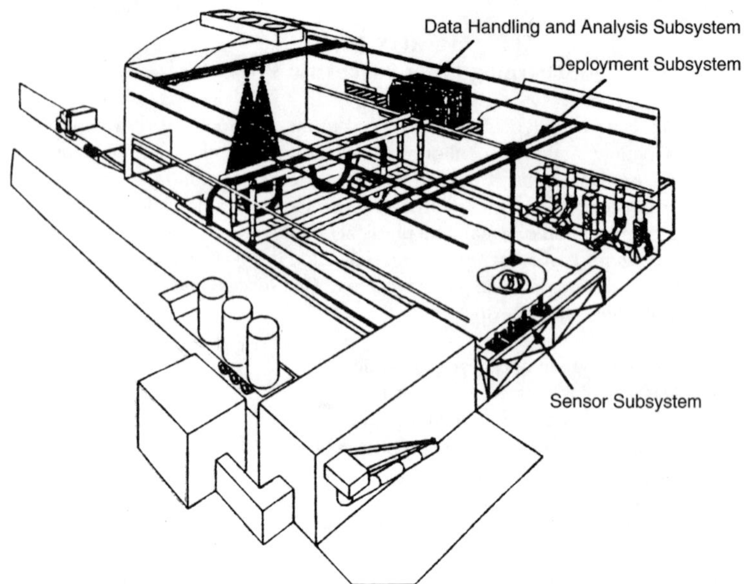

FIGURE 5-3 Dig-face characterization system. SOURCE: Josten et al., 1995.

Representative Successful Technologies for Remediation of Metals and Radionuclides

Representative successful SCFA achievements in developing technologies for remediation of metals and radionuclides include in situ redox manipulation systems for chromium attenuation at Hanford, bottom barriers for waste containment at the Idaho National Engineering and Environmental Laboratory (INEEL), and the site characterization and analysis penetrometer system (SCAPS) for characterization of subsurface environments in a wide range of settings.

In Situ Redox Manipulation

SCFA has provided funding for the creation and operation of a permeable treatment zone for remediation of Cr(VI) in the contaminated aquifer at Hanford by in situ redox manipulation (ISRM) (see Box 5-2 and Figure 5-4). The demonstration of this system at Hanford has shown that this process is relatively inexpensive: it is comparable in cost to an impermeable barrier and is able to provide an overall cost savings of approximately 60 percent compared to a pump-and-treat system for the preven-

BOX 5-2
In Situ Redox Manipulation for Remediation of Chromium-Contaminated Groundwater at Hanford, Washington

Contamination Source. Hexavalent chromium, Cr(VI), in the form of sodium dichromate was used as an anticorrosion agent in the cooling water for the nine nuclear reactors at Hanford. Large volumes of reactor coolant water, along with liquid wastes from other reactor operations that also contained significant quantities of Cr(VI), were discharged to retention basins for ultimate disposal in the Columbia River through outfall pipelines. Discharge of these liquids created contaminant plumes in groundwater that are flowing toward and entering the Columbia River.

Procedure. The ISRM technology, being developed by Hanford researchers with funding from SCFA, is based on creation of a permeable subsurface treatment zone for remediating redox-sensitive contaminants in groundwater. The treatment zone is created downgradient of the contaminant plume or contaminant source through the reduction of ferric iron, Fe(III), to ferrous iron, Fe(II), within the silt and clay minerals of the aquifer sediments. Comparative laboratory-scale batch studies with sulfite, thiosulfate, hydroxylamine, and dithionite under anoxic conditions established that dithionite was the most effective reducing agent for the structural ferric iron found in the silt and clay fractions of Hanford sediments. Similar experiments were used to identify a pH buffer for use with dithionite. The reagent used is 0.4 M K_2CO_3 + 0.04 M $KHCO_3$ + 0.1 M $Na_2S_2O_4$. Carbonate was selected for the buffer because it has no toxic properties.

The permeable treatment zone is created using a push-pull technique. The reagent, buffers, and tracers are pumped into the aquifer (injection phase), allowed to react for a period determined by laboratory and field demonstration experiments (reaction-drift phase), and then pumped back out (withdrawal phase). During the reaction-drift phase, the dithionite ion dissociates into sulfoxyl radicals that either reduce ferric iron to ferrous iron or disproportionate into thiosulfate and bisulfate. After the aquifer sediments are reduced, soluble reagents and reaction products are removed. The reduced iron in the soil acts as a permeable treatment barrier by reducing chromate to insoluble chromium hydroxide. The lifetime of the permeable treatment barrier depends on the pollutant concentration, but the primary driver for Fe(II) depletion is the concentration of oxygen in groundwater.

Because of the proximity of the site location to the Columbia River, a contingency plan has been developed in the event that dissolved oxygen concentrations are severely reduced in the groundwater entering the river. The contingency plan involves pumping groundwater from the injection-withdrawal wells and the use of downgradient monitoring wells to rapidly reoxygenate the reduced zone. Pumping will stop once the dissolved oxygen concentratons at the site are back to preemplacement levels.

Cost Effectiveness. An independent Los Alamos National Laboratory assessment concluded that ISRM costs 62 percent less than a pump-and-treat system for prevention of chromium movement in an unconfined aquifer under the small-scale conditions considered.

Project History. In addition to laboratory-scale demonstrations, three field experiments have been conducted: (1) a full-scale bromide tracer experiment; (2) a small-scale "mini" dithionite injection-withdrawal experiment; and (3) a full-scale dithionite injection-withdrawal experiment. The permeable treatment barrier is now slated for full-scale use in cleanup of Cr(VI) at Hanford.

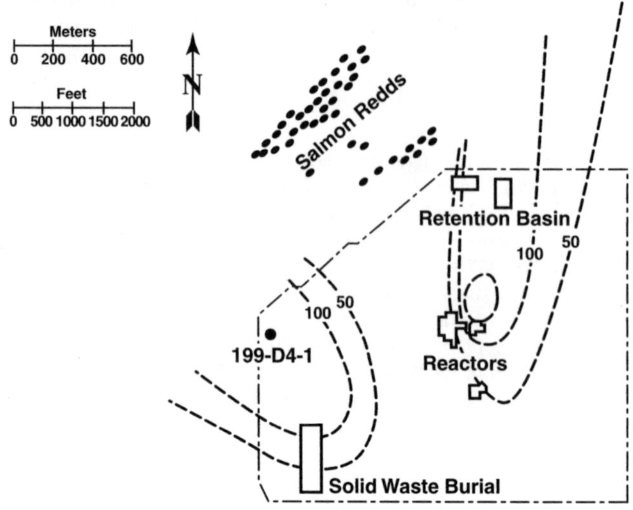

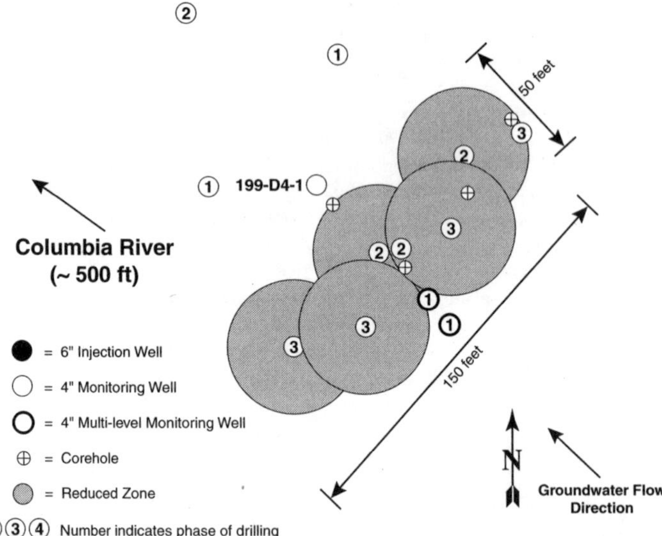

FIGURE 5-4 Plan of the in situ redox manipulation system at Hanford. SOURCE: Fruchter, 1997.

tion of chromium migration in an uncontrolled aquifer, with a long-term prognosis of additional cost savings in the future (Cummings and Booth, 1996; Civil Engineering, 1998). In the demonstration, analysis of water withdrawn downgradient of the treatment zone indicated that all trace metals, including arsenic, lead, and chromium, were below the 0.1-ppm (part per million) detection limit. Because this is an in situ technique, it reduces the risk of exposure to contamination and eliminates the need for permanent external pumping and treatment systems.

Buried Waste Containment System

Based on needs identified by DOE field site managers, SCFA has recognized the need for technology that would allow placement of a continuous barrier under and around buried wastes and has provided funding for the development of such barriers With support from SCFA, INEEL is developing the buried waste containment system (BWCS), which places a continuous, seamless barrier under and around buried waste (see Figure 5-5). This system is applicable to buried wastes containing metals and radionuclides, as well as other types of contaminants. Using an innovative, positive-displacement grouting technique, the system excavates the material under and around the buried waste and simultaneously replaces it with a barrier material to contain the waste. The BWCS design includes equipment to verify and monitor barrier integrity, both during placement and over the long term.

The BWCS was jointly developed by INEEL and R. A. Hanson Company (RAHCO) via a cooperative research and development agreement (CRADA), with the intent to develop a licensing agreement with RAHCO International. Results include a conceptual design, preliminary plan for bench-scale testing, identification of verification and monitoring technologies, and preliminary barrier material content. Two patents have been filed, and a life-cycle development plan has been written. The ability of this technology to address a common problem in the DOE complex and the availability of a contractor to provide the technology create opportunities to pursue its deployment at other DOE sites at which implementation of other technologies, such as excavation, is not feasible.

Site Characterization and Analysis Penetrometer System

Recognizing the need at DOE sites for technologies that can allow real-time on-site analysis of the subsurface, SCFA has contributed funds toward the development of SCAPS by a government consortium. SCAPS allows rapid characterization of subsurface environments using push probes for investigation and sampling (see Figure 5-6). It provides in situ

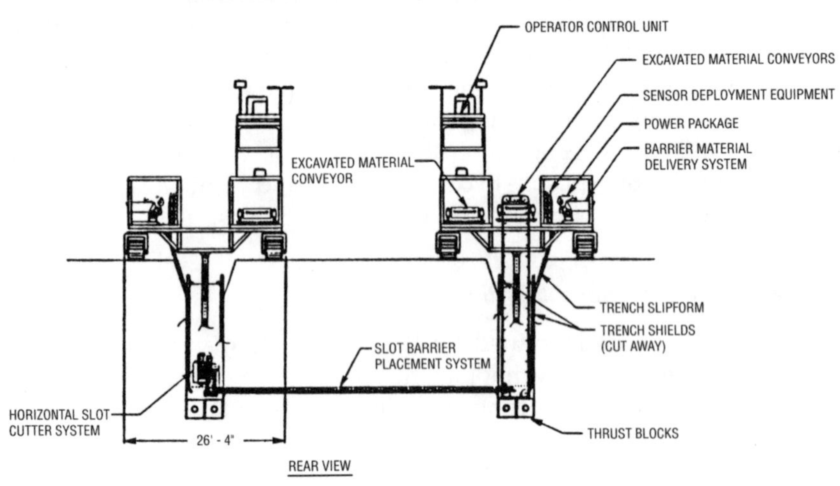

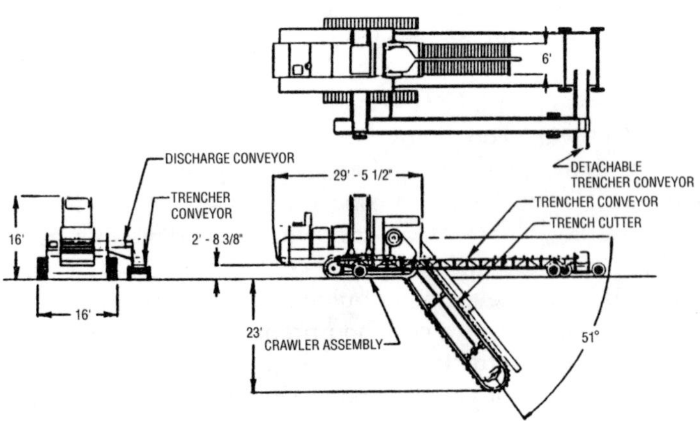

FIGURE 5-5 Buried waste containment system being developed by INEEL with funding from SCFA. SOURCE: Crocker, 1997.

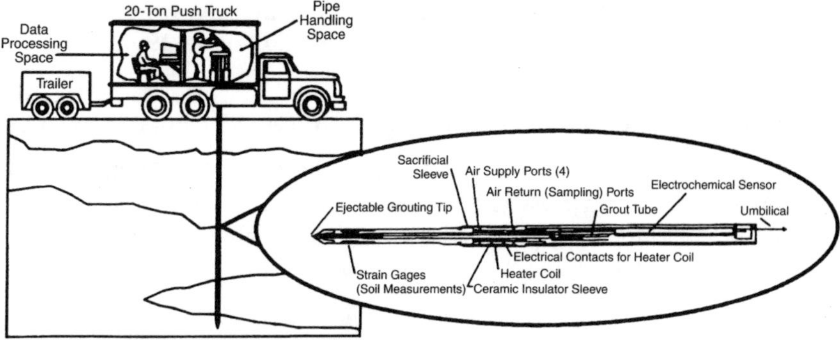

FIGURE 5-6 Site characterization and analysis penetrometer system, developed with partial support from SCFA. SOURCE: U.S. Army Corps of Engineers Waterways Experiment Station, undated.

measurements of geophysical and physical properties of soils and stratigraphic units, as well as contaminant concentrations, at a site without extensive use of drills and monitoring wells. SCAPS can also collect soil and water data to better define zones of contamination, enabling more accurate placement of remediation systems and monitoring wells. It applies to sites containing a wide variety of contaminants, including metals and radionuclides.

Originally developed by the Waterways Experiment Station with sponsorship from the U.S. Army Environmental Center and later further developed in collaboration with the Navy and DOE, SCAPS consists of a 20-ton truck equipped to force a cone penetrometer sensor probe into the ground, a data acquisition-processing room, and a hydraulic ram-rod handling room. SCAPS probes have multisensor capabilities with an onboard system providing real-time data acquisition, processing, and storage; an electronic signal processing equipment package; and a networked postprocessing computer system for three-dimensional visualization of soil stratigraphy and contaminant plumes. A mobile laboratory truck, equipped with a field-portable ion trap mass spectrometer and/or gas chromatograph, accompanies the SCAPS truck for real-time on-site analysis of analyte vapor samples collected by SCAPS in situ samplers. A variety of sensors and samplers can be deployed with SCAPS to detect a range of contaminants, from metals and radionuclides to volatile organic compounds.

SCAPS technology is being used by the Army Corps of Engineers, Department of Defense, other government agencies, and the private sector, as well as by DOE, through licensing and CRADA agreements. Use of SCAPS site characterization and monitoring technologies typically provides cost savings of 25 to 50 percent per site compared to conventional drilling and sampling techniques (Ballard and Cullinane, 1997).

Representative Successful Technologies for Remediation of DNAPLs

Representative successful SCFA projects for developing dense nonaqueous-phase liquid (DNAPL) remediation technologies include dynamic underground stripping (DUS), thermally enhanced vapor extraction (TEVES), and Lasagna®.

Dynamic Underground Stripping

DUS is technically a highly effective system for removing free-phase DNAPL. It addresses a commonly identified problem in the DOE complex for which cost-effective solutions are extremely limited. The DUS system, which is being developed with partial support from SCFA, combines three technologies:

1. steam injection at the periphery of a contaminated area to drive contaminants to centrally located vacuum extraction locations;
2. electrical heating of less permeable soils; and
3. underground imaging (using electrical resistance tomography) to delineate heated areas.

Surrounding an underground plume with injection wells and electrically heating clay-rich soil layers, while sandy layers, are flooded with steam, volatilize contaminants, which the steam then carries to extraction wells. The steam is condensed, extracted, and treated above ground. Water condensed from the steam is reinjected underground after the contaminants are removed. The process is capable of removing free DNAPL product. Time savings of an order of magnitude, which translate into considerable cost savings, are considered possible compared to pump-and-treat technology for a broad range of DNAPL contaminants (Aines, 1997).

The original demonstration of DUS, conducted in 1992-1993 by Lawrence Livermore National Laboratory (LLNL), evaluated the effectiveness of this technology for cleanup of a gasoline spill site (see Box 5-3). LLNL researchers compared results to those from a pump-and-treat system and determined that the potential cost savings of applying DUS, instead of a pump-and-treat system, at the same site in the future would be $4 million, when benefits of lessons learned and reduced costs for deployment without research-oriented activities are taken into account

BOX 5-3
Dynamic Underground Stripping for Remediation of Gasoline-Contaminated Groundwater at LLNL

Contamination Source. An estimated 65 m^3 (17,000 gallons) of leaded gasoline leaked from underground storage tanks between 1952 and 1979 at LLNL, at a site now called the Gasoline Spill Area.

Procedure. DUS combines steam injection and electrical heating to drive non-aqueous-phase liquid contaminants from the subsurface. In this full-scale demonstration, six wells combining steam injection and electrical heating, three wells using electrical heating alone, and one vacuum extraction system were used to clean up the fuel hydrocarbons. Well characteristics were as follows:

- steam injection-electrical heating wells: 44.2 m (145 ft) deep, 10-cm (4-in) diameter, screened in upper and lower steam zones;
- electrical heating wells: 36.6 m (120 ft deep), 5-cm (2-in.) diameter; and
- groundwater and vapor extraction well, 47.2 m (155 ft) deep, 20-cm (8-in) diameter.

Extracted water was processed through a heat exchanger, oil-water separators, filters, ultraviolet light and hydrogen peroxide treatment units, air strippers, and granular activated carbon filters. Extracted vapors were processed through a heat exchanger, demister, and internal combustion engine.

Results. The demonstration resulted in the removal of more than 29 m^3 (7,600 gallons) of gasoline, mostly in the vapor stream rather than in the extracted groundwater.

Cost Effectiveness. Researchers estimated that potential cost savings from the use of DUS, rather than a pump-and-treat system, for full-scale treatment of this site are $4 million.

Project History. The demonstration began in November 1992 and ended in December 1993.

SOURCE: Federal Remediation Technologies Roundtable, 1995.

(Federal Remediation Technologies Roundtable, 1997). Overall program costs for the field demonstration were $1.7 million for before-treatment costs and $5.4 million for treatment activities.

The first full-scale commercial DUS application is ongoing at Southern California Edison's Visalia Pole Yard. The project involves a partnership among LLNL, Southern California Edison, and SteamTech Environmental Services as the licensee. The site is contaminated with creosote. The use of DUS is expected to allow site closure in five years at a cost savings to the company of $30 million compared with a conventional pump-and-treat remedy (Aines, 1997).

Thermally Enhanced Vapor Extraction

TEVES is one of the technology development projects that SCFA is supporting to address the difficulty of removing contaminants with low volatility from low-permeability soils. The technology couples soil heating by resistive and dielectric (radio-frequency) methods with vacuum vapor extraction (see Figure 5-7). Although the use of electrical heating techniques for the recovery of volatile and semivolatile liquids from porous media is not new, the use of resistance heating for in situ recovery is more recent. In TEVES, three rows of electrodes are placed through a contaminated zone with the center electrodes connected to the energy input (excitor) and the two exterior rows serving as a grounding system to help contain the input energy to the treatment zone. Two wells providing for soil vapor extraction and also containing electrodes are installed as part of the excitor array. A vacuum blower and off-gas treatment system are provided for the removal of the heated soil contaminants.

A field demonstration at Sandia National Laboratories evaluated the application of TEVES on an old disposal pit containing a complex mixture of organic chemicals, oils, and containerized wastes (Sandia National Laboratories, undated). Process monitoring systems included automated vapor sampling and analysis of the extracted contaminants and subsurface pressure to monitor vapor capture in the treatment zone. Resistive heating for 30 days increased soil temperature to 83°C over the entire treatment volume. Contaminant concentration removal in the gas phase increased by 400 percent compared to extraction at ambient temperature. Subsequent cooling to ambient temperature and radio-frequency heating for 30 days raised the average soil temperature to 112°C, with a contaminant concentration increase of 500 to 1,000 percent over baseline.

TEVES also has been applied to pilot-scale cleanup of trichloroethylene (TCE) and a gasoline spill at LLNL (in 1992 and 1993). In the initial LLNL investigation, a three-phase 400-V power source heated a region about 7 m in diameter and 4 m thick with six electrodes placed symmetrically around the periphery, with an extraction well in the center of the zone. The electrical heating ran for 47 days. The temperature in the middle of the pattern increased from 19° to 44°C and to 55°C after heating was discontinued. Vapor TCE concentrations increased by a factor of two compared to stable rates obtained by vacuum extraction alone; vapor concentrations decreased rapidly near the end of electrical heating (Udell, undated).

Coinciding with the final phase of electrical heating at LLNL, Pacific Northwest Laboratories (PNL) used electrical heating at Savannah River to remove perchloroethylene (PCE), TCE, and trichloroethane (TCA) from low-permeability clays in the vadose zone (see Box 5-4). On initiation of

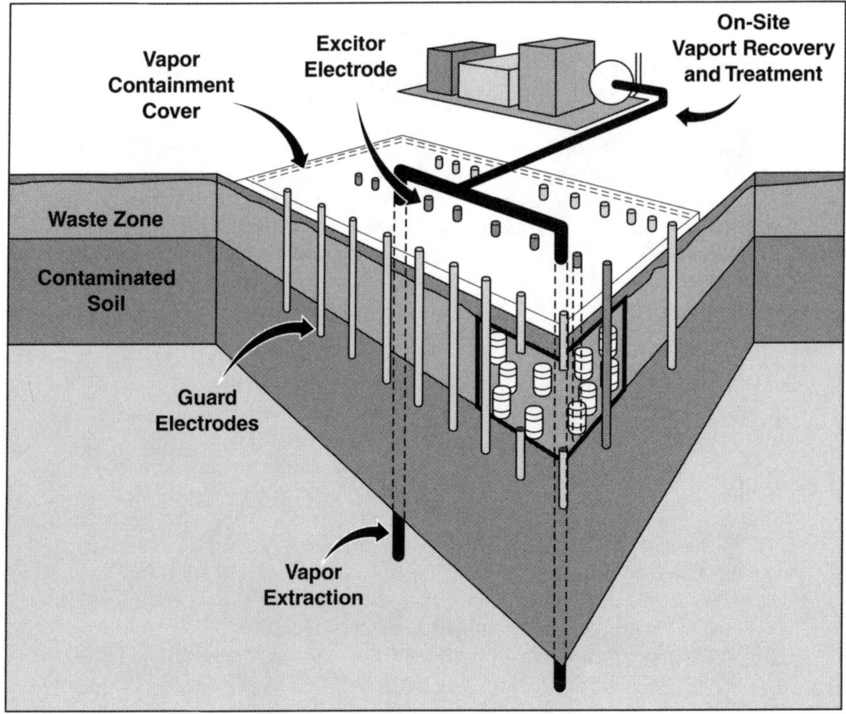

FIGURE 5-7 Thermal-enhanced vapor extraction system, being developed at Sandia National Laboratories with funding from SCFA. SOURCE: Sandia National Laboratories, undated.

the electrical heating, slight increases in contaminant recovery rates in the air leaving the treatment condenser, beyond those predicted for soil vapor extraction alone, were observed, although the location of the demonstration inside a larger contaminated zone obfuscated the vapor concentration results (Udell, undated). Soil concentrations decreased on average by more than 99 percent inside the pattern and more than 95 percent outside the pattern in heated zones (Udell, undated).

Based on reported results, electrical resistance heating combined with vapor extraction for in situ cleanup of DNAPL contaminants found both above and below the water table in low-permeability media is a promising technique. With proper design and operation, this remediation method is expected to be relatively rapid, robust, and predictable. The cost to remediate a site would depend on the required number of extraction and electrode wells, access to adequate line power, and fluid treatment requirements.

BOX 5-4
Electrical Heating for Treatment of Solvent-Contaminated Soil at the Savannah River Site

Contamination Source. From 1958 until 1985, process wastewater from metal manufacturing operations at the Savannah River Site was disposed of in an unlined settling basin. The wastewater contained TCE, PCE, and TCA, which subsequently migrated to the soil and groundwater beneath the settling basins.

Procedure. The heating system used in this demonstration was created by splitting conventional three-phase electricity into six separate phases, each of which was delivered to a different electrode. The six electrodes were set into a hexagonal pattern, 9.1 m (30 ft) in diameter. Moisture was maintained at the electrodes by adding 4 to 8 liters per hour (1 to 2 gallons per hour) of a 500-mg/liter sodium chloride solution to each electrode. A vapor extraction well was located in the center of the hexagon to withdraw contaminants volatilized by the application of heat. Power was applied to the electrodes for a total of 25 days.

Results. After eight days of heating, the soil temperature rose to 100°C; the temperature stabilized at 100 to 110°C for the remaining 17 days of the demonstration. The system removed 180 kg of PCE and 23 kg of TCE. Median PCE removal was 99.9 percent. Researchers estimated that cleanup of the site using this method would require 5 years, compared to 50 years for soil vapor extraction alone. Operating difficulties that required adjustments of the system during the test period included drying out of the electrodes and shorting of the thermocouples.

Cost Effectiveness. Researchers estimated the cost of this system at $\$110/m^2$ ($\$86/yd^3$) of soil treated, compared to an estimated cost of $\$753/m^3$ ($\$576/yd^3$) for soil vapor extraction.

Project History. This demonstration was conducted from October 1993 through January 1994.

SOURCE: Federal Remediation Technologies Roundtable, 1995.

Lasagna® Soil Remediation

Another technology development project that SCFA is helping to support to address the problem of cleanup of low-permeability zones is the Lasagna® process. This system couples electrokinetics with in situ treatment zones. The process was developed by a consortium including Monsanto, E. I. DuPont de Nemours & Co., and General Electric, with participation from DOE and EPA. As indicated in Figure 5-8, the name "Lasagna" derives from the original concept of alternating horizontal layers of electrodes and treatment zones, although actual tests to date have used a vertical configuration. The process is especially suited to sites with low-permeability soils because electroosmosis can move water faster and more uniformly through such soils than hydraulic methods and because

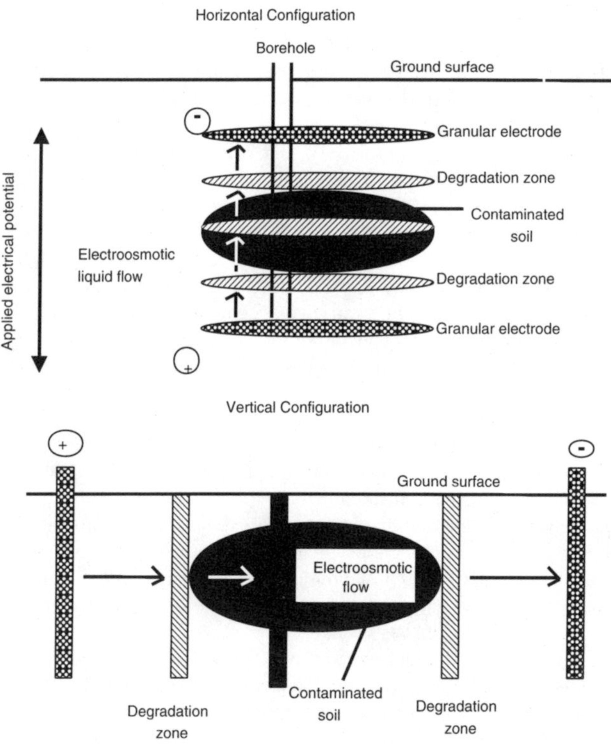

FIGURE 5-8 Lasagna® process, developed with partial support from SCFA. SOURCE: Ho, 1997.

electrokinetics can move contaminants in soil pore water to treatment zones, where they can be captured or transformed. Major features of the technology are

- electrodes, energized by direct current, that heat the soil and cause water and soluble contaminants to move through the treatment layers;
- treatment zones containing reagents that transform the soluble organic contaminants or adsorb contaminants for immobilization or subsequent removal and disposal; and
- a water management system to recycle the water that accumulates at the cathode (high pH) back to the anode (low pH) for acid-base neutralization or, alternatively, periodic reversal of electrode polarity to reverse electroosmotic flow and neutralize pH.

BOX 5-5
Cleanup of TCE in Soil Using the Lasagna® Method at the Paducah Gaseous Diffusion Plant

Contaminated Site Characteristics. The site used in this field demonstration was a 4.6 m × 3.0 m (15 ft × 10 ft) square plot at the Paducah Gaseous Diffusion Plant cylinder drop-test pad area. Soil at the site contained from less than 1 to 1,500 μg/g of TCE. The soil consists mostly of clay, with a porosity of 0.4.

Procedure. Two 4.6-m (15 ft) vertical electrodes were emplaced 3.0 m (10 ft) apart to a depth of approximately 4.6 m (15 ft). Between the electrodes, four rows of wicks filled with granular activated carbon were emplaced approximately 0.6 m (2 ft) apart. Voltage was applied to the electrodes for 120 days at a current of 40 amperes and a voltage gradient of 0.45 to 0.35 V/cm. The induced flow rate averaged about 4 liters per hour, resulting in about three pore volumes of water being circulated during the four-month operating period.

Results. TCE removal, based on soil core analyses, averaged 98.4 percent, with final TCE concentrations generally below 1 mg/kg soil. Higher residuals were found at the base of the test zone, indicating that contamination extended to greater depth. Approximately 50 percent of the estimated original mass of TCE was captured on the carbon wicks. Several core samples yielded calculated TCE concentrations in the pore water above the TCE solubility limit, suggesting that DNAPL was present at the start of the test. Residual values were low in these areas, suggesting that the process was effective where DNAPL was present.

Cost-Effectiveness. Although no data were provided on the capital or operating costs for this demonstration, the industry-government consortium responsible for developing Lasagna® has estimated costs based on data from this demonstration, a later demonstration, and a paper study of a full-scale cleanup operation. Costs (excluding those for sampling and oversight) ranged from approximately $\$160/m^3$ ($\$120/yd^3$) of soil under optimal conditions to nearly $\$340/m^3$ ($\$260/yd^3$) under difficult conditions. These costs were determined based on a 18 m x 30 m (60 ft x 100 ft) treatment zone with a depth of either 4.6 m (15 ft) or 14 m (45 ft).

Project History. This demonstration operated from January though May 1995 and was followed with a larger field test.

SOURCES: Federal Remediation Technologies Roundtable, 1995; Monsanto Company, 1998.

The first field test of Lasagna® was conducted in 1995 at DOE's Paducah Gaseous Diffusion Plant in Kentucky (see Box 5-5). Based on promising results from this first test, a larger-scale field test was conducted at Paducah in 1996-1997 (Monsanto Company, 1998). This test used two electrodes, each 9.1 m (30 ft) long and 14 m (45 ft) deep, spaced 6.4 m (21 ft) apart. Three treatment zones containing zero-valent iron were installed at 2.1, 3.7, and 4.3 m (7, 12, and 14 ft) from the anode. The system was operated for one year, resulting in circulation of about 2.5

pore volumes of water. Soil temperature was raised to over 60°C throughout the test volume, reaching 80°C in the center. TCE removal efficiencies ranged from 41.5 to 99.7 percent. The technology performed as effectively in areas believed to contain DNAPL as in areas that did not.

The use of treatment zones for in situ destruction of contaminants gives Lasagna® a competitive advantage over other electrokinetic methods that extract contaminants for above ground treatment or disposal. The implementation cost for Lasagna® in the initial studies was estimated by DuPont at $100-$120/m^3 ($80-$90/yd^3) for remediation in one year and $65-$78/m^3 ($50-$60/yd^3) if three-year remediation was allowed (DOE, 1996). Comparable preliminary estimates for the second field test were $78-$92/m^3 ($60-$70/yd^3) (one year) and $52-$65/m^3 ($40-$50/yd^3) (three years).[4]

CONCLUSIONS

DOE is not alone in facing resistance to the use of innovative technologies for cleaning up contaminated soil and groundwater at its installations. Use of innovative remediation technologies is also quite limited in private-sector cleanup of major contaminated sites. At both DOE installations and private-sector sites, a primary barrier to the use of innovative remediation technologies is lack of demand for such technologies by end users.

SCFA's potential for progress also has been limited considerably by its small and continually declining budget. The 1998 budget of approximately $10 million is less than half the cost of cleaning up one typical CERCLA site. DOE managers will have to reassess whether this budget adequately reflects the level of priority that should be given to developing new groundwater and soil remediation technologies.

The committee believes that SCFA has an important mission to fulfill in continuing development work on innovative remediation technologies, especially those for cleaning up metals, radionuclides, and DNAPLs. The technical solutions for these types of contamination problems are generally not adequate or are excessively costly. Key areas of concern for ensuring the success of future SCFA technology development efforts are as follows:

- **The limited SCFA budget.** SCFA's budget has been reduced so much that it is unlikely SCFA can have a significant impact on the development of innovative remediation technologies. The budget was cut from a

[4] Cost estimates include direct costs of the technology only.

1994 level of $82 million to a 1998 level of $15 million, which includes a $5 million congressional earmark, leaving an effective budget of $10 million.

- **The lack of incentives and cost control for site cleanup**. Lack of sufficient incentives from DOE headquarters for prompt and cost-effective cleanup of DOE sites is a critical barrier to SCFA's successful development and deployment of innovative remediation technologies. Local control of technology selection does not provide the broad perspective needed for maximizing returns on limited DOE funds.
- **The high perceived risk of initial technology deployment.** Contractors, as well as regulators, at DOE installations can be reluctant to accept the full consequences of failure should a potentially cost-effective innovative remediation technology fail to perform as predicted and thus will tend to choose conventional remediation technologies over innovative ones.
- **Insufficient data on full-scale deployment of SCFA technologies**. Data on applications of innovative remediation technologies at DOE sites are currently inadequate to determine the full extent of the use of SCFA technologies in site cleanup.
- **Need for greater collaboration with leaders in the field of remediation technology development**. SCFA has taken credit for the development of a number of technologies for which sufficient research and development efforts already had occurred in the private sector. This overlap suggests lack of a sufficient partnering strategy between SCFA and external technology developers. It also suggests lack of sufficient expertise among SCFA staff with respect to technologies developed outside SCFA.
- **Need for greater involvement of technology end users in the SCFA program.** Despite SCFA's formation of STCGs, the field personnel who are the ultimate customers for SCFA's technologies still are not adequately involved in setting overall program direction and planning individual technology development projects.
- **Need for multisite applications of SCFA technologies.** Fewer than one-third of SCFA technologies have been deployed at more than one facility, and fewer than 20 percent have been deployed at more than two facilities.
- **Need for more work on in situ remediation technologies.** Fewer than one-third of SCFA technologies address the need for in situ remediation of contaminants in soil and groundwater. Development of in situ remediation technologies may not be receiving appropriate priority.

REFERENCES

Aines, R. 1997. Results from Visalia: Rapid thermal cleanup of dense nonaqueous-phase liquids. Presentation to the National Research Council Committee on Technologies for Cleanup of Subsurface Contaminants in the DOE Weapons Complex, Third Meeting, Livermore, Calif., December 15-17.

Ballard, J. H., and M. J. Cullinane, Jr. 1997. Tri-Service Site Characterization and Analysis Penetrometer System (SCAPS) Technology Verification and Transition. Vicksburg, Miss.: U.S. Army Corps of Engineer Waterways Experiment Station.

Baum, J. S. 1998a. SCFA processes for setting project funding priorities. Presentation to the National Research Council Committee on Technologies for Cleanup of Subsurface Contaminants in the DOE Weapons Complex, Fifth Meeting, Augusta, Ga. February 18-20.

Baum, J. S. 1998b. Written memorandum in answer to questions submitted by the Committee on Technologies for Cleanup of Subsurface Contaminants in the DOE Weapons Complex. Aitken, S.C.: DOE, Savannah River Site.

Brown, R. A., W. Mahaffey, and R. D. Norris. 1993. In situ bioremediation: The state of the practice. Pp. 121-135 in In Situ Bioremediation: When Does It Work? Washington, D.C.: National Academy Press.

Civil Engineering. 1998. Chemical barrier remediates groundwater. Civil Engineering (November):14.

Cooper, R. K. 1993. Winning at New Products. New York: Perseus Press.

Crocker, T. 1997. Buried waste containment system: Results of recent research. Presentation to the National Research Council Committee on Technologies for Cleanup of Subsurface Contaminants in the DOE Weapons Complex, Second Meeting, Woods Hole, Mass., September 22-23.

Cummings, M., and S. R. Booth. 1996. Cost Effectiveness of In Situ Redox Manipulation for Remediation of Chromium-Contaminated Groundwater. Los Alamos, N.M.: Los Alamos National Laboratory.

DOE (Department of Energy). 1996. Innovative Technology Summary Report: Lasagna® Soil Remediation. DOE/EM-0308. Washington, D.C.: DOE.

EPA (Environmental Protection Agency). 1996. Innovative Treatment Technologies: Annual Status Report (Eighth Edition). EPA-542-R-96-010. Washington, D.C.: EPA, Office of Solid Waste and Emergency Response.

EPA. 1997. Superfund Public Information System. Washington, D.C.: EPA, Office of Emergency and Remedial Response. http://www.epa.gov/Superfund.

Federal Remediation Technologies Roundtable. 1995. Abstracts of Remediation Case Studies. EPA-542-R-95-001. Washington, D.C.: Environmental Protection Agency.

Federal Remediation Technologies Roundtable. 1997. Abstracts of Remediation Case Studies. EPA-542-R-97-010. Washington, D.C.: Environmental Protection Agency.

Fruchter, J. S. 1997. In situ redox manipulation: Results of recent research. Presentation to the National Research Council Committee on Technologies for Cleanup of Subsurface Contaminants in the DOE Weapons Complex, Second Meeting, Woods Hole, Mass., September 22-23.

GAO (U.S. General Accounting Office). 1992. Cleanup Technology: Better Management for DOE's Technology Development Program. GAO/RCED-92-145. Washington, D.C.: GAO.

GAO. 1994a. Department of Energy: Management Changes Needed to Expand Use of Innovative Cleanup Technologies. GAO/RCED-94-205. Washington, D.C.: GAO.

GAO. 1994b. Nuclear Cleanup: Difficulties in Coordinating Activities Under Two Environmental Laws. GAO/RCED-95-66. Washington, D.C.: GAO.

GAO. 1996a. Energy Management: Technology Development Program Taking Action to Address Problems. GAO/RCED-96-184. Washington, D.C.: GAO.

GAO. 1996b. Superfund: How States Establish and Apply Environmental Standards When Cleaning up Sites. GAO/RCED-96-70. Washington, D.C.: GAO.

GAO. 1997. Department of Energy: Funding and Workforce Reduced, but Spending Remains Stable. GAO/RCED-97-96. Washington, D.C.: GAO.

GAO. 1998a. DOE: Alternative Financing and Contracting Strategies for Cleanup Projects. Washington, D.C.: GAO.

GAO. 1998b. Nuclear Waste: Further Actions Needed to Increase the Use of Innovative Cleanup Technologies. GAO/RCED-98-249. Washington, D.C.: GAO.

Guerrero, P. F. 1997. Federal hazardous waste sites: Opportunities for more cost-effective cleanups. Testimony before the Subcommittee on Superfund, Waste Control, and Risk Assessment, Committee on Environment and Public Works, U.S. Senate, May 9, 1995. Washington, D.C.: U.S. General Accounting Office.

Hill, G. R., C. H. Sink, and D. A. Lynn. 1997. Three Tiers and Six Gates: Implementing Streamlined Regulatory Review and Acceptance of Innovative Environmental Technologies Draft for Review. Norcross, Ga.: Southern States Energy Board.

Ho, S. V. 1997. Scale-up aspects of the Lasagna® process for in situ soil decontamination. Journal of Hazardous Materials 55:39-60.

Josten, N. E., R. J. Gehrke, and M. V. Carpenter. 1995. Dig-face Monitoring During Excavation of a Radioactive Plume at Mound Laboratory, Ohio. INEL-95/0633. Idaho Falls, Id.: Idaho National Engineering and Environmental Laboratory.

MacDonald, J. A. 1997. Hard times for innovative cleanup technology: What can be done to remove market barriers to new groundwater and soil remediation technologies? Environmental Science and Technology 31(12):560A-563A.

Mintz, J. 1997. Lockheed Martin mired in toxic mess cleanup: Company failing to fulfill government contract. Washington Post (October 5):A14.

Monsanto Company. 1998. Rapid Commercialization Initiative (RCI) Final Report for an Integrated In-Situ Remediation Technology (Lasagna®). Morgantown, W.V.: Morgantown Energy Technology Center. http://www.rtdf.org/lastechp.htm.

NRC (National Research Council). 1997a. Innovations in Ground Water and Soil Cleanup: From Concept to Commercialization. Washington, D.C.: National Academy Press.

NRC. 1997b. Peer Review in the Department of Energy—Office of Science and Technology: Interim Report. Washington, D.C.: National Academy Press.

NRC. In review. Decision Making Related to the U.S. Department of Energy's Environmental Management Office of Science and Technology. Washington, D.C.: National Academy Press.

Nemeth, K. J., C. H. Sink, G. R. Hill, and A. Rappazzo. 1997. Strategy for regulatory review and acceptance of innovative environmental technology. Presented at X-Change 97: The Global D&D Marketplace, Miami, Fla., December 1-5.

Rezendes, V. S. 1997. Cleanup technology: DOE's program to develop new technologies for environmental cleanup. Testimony before the Subcommittee on Oversight and Investigations, Committee on Commerce, U.S. House of Representatives, May 7, 1997. GAO/T-RCED-97-161. Washington, D.C.: U.S. General Accounting Office.

Sandia National Laboratories. Undated. Thermal Enhanced Vapor Extraction System. Albuquerque, N.M.: Sandia National Laboratories.

Udell, K. S. Undated. Thermal Treatment of Low Permeability Soils Using Electrical Resistance Heating. Berkeley, Calif.: Berkeley Environmental Restoration Center, College of Engineering, University of California.

U.S. Army Corps of Engineers. 1997. U.S. Army Corps of Engineers Peer Review of DOE Office of Science and Technology (EM-50) Cost Savings Calculations. Washington, D.C.: U.S. Army Corps of Engineers.

U.S. Army Corps of Engineers Waterways Experiment Station. Undated. Tri-Service Site Characterization and Analysis Penetrometer System (SCAPS): Technology Development/Transition. Vicksburg, Miss.: Waterways Experiment Station.

6

Findings and Recommendations

The Department of Energy (DOE) will be unable to meet all of the applicable federal and state regulations for cleanup of contaminated groundwater and soil at its facilities with existing technologies. The most intractable problems involve dense nonaqueous-phase liquids (DNAPLs), metals, and radionuclides. The Subsurface Contaminants Focus Area (SCFA) within DOE's Office of Science and Technology (OST) is charged with developing innovative, cost-effective technologies to address these intransigent problems. As described in this report, SCFA has achieved a number of successes, but its progress has been limited by budget and programmatic problems, both in SCFA and in the DOE environmental restoration program as a whole.

The Committee on Technologies for Cleanup of Subsurface Contaminants in the DOE Weapons Complex, at the request of DOE and in the course of preparing this report, developed a series of findings and recommendations for improving SCFA's technology development program. The major findings and recommendations are presented in this chapter and are based on the analyses provided in Chapters 2 through 5. They are organized into four categories: (1) technology development priorities, (2) overall program direction, (3) barriers to deployment, and (4) budget limitations.

SETTING TECHNOLOGY DEVELOPMENT PRIORITIES

Finding. Because many metal, radionuclide, and DNAPL remediation technologies are in their early stages of development, relatively rapid progress should be expected with continuing work by consulting firms, private industries, and academic and government laboratories involved in remediation technology development, as well as by SCFA.

- ***Recommendation:*** SCFA should develop and maintain a system for updating technology evaluations for remediation of metals, radionuclides, and DNAPLs. In order to avoid duplicating the work of others, SCFA needs to keep apprised of and selectively use results from remediation technology development projects by outside organizations.

Finding. Fewer than one-third of SCFA technologies address the need for in situ remediation of contaminants in groundwater and soil. Development of in situ remediation technologies appears not to be receiving appropriate priority.

- ***Recommendation:*** In situ remediation technologies should receive a higher priority in SCFA because of their potential to reduce exposure risks and costs.

Finding. Promising technologies for remediation of metals and radionuclides include ion exchange systems and electrokinetic technologies. Ion exchange methods are simple and potentially effective for use in in situ barriers for metals and radionuclides, particularly if more selective ion exchange media can be developed. Electrokinetic technologies appear promising for extraction of metal and radionuclide contaminants from fine-grained media, but additional field demonstrations are necessary to establish performance under field conditions encountered at DOE sites.

- ***Recommendation:*** SCFA should consider funding work on the development of selective ion exchange media for use in reactive barriers. SCFA also should fund additional field demonstrations of electrokinetic systems, building on private-sector and overseas tests and focusing on metals and radionuclides unique to the weapons complex.

Finding. Although it is unlikely that any remediation technology will restore every portion of an aquifer contaminated with DNAPLs to baseline standards, some emerging technologies have demonstrated applicability for removing significant amounts of contaminant mass from DNAPL source zones. These technologies include systems using surfactants, cosolvents, steam and other forms of heat, soil vapor extraction (including thermal enhancements), air sparging, and in situ oxidation. In addition, several technologies have potential applicability primarily for aqueous-phase contamination dissolved from DNAPLs. These technologies include various bioremediation and reactive barrier wall systems, in addition to conventional pump-and-treat systems.

- *Recommendation:* Although the potential of the DNAPL remediation technologies listed above has been demonstrated, SCFA should fund additional, carefully controlled tests in conjunction with external technology developers (where appropriate) to provide cost and performance data to facilitate application of these technologies. There is much uncertainty in technology evaluations due to the limited amount of high-quality data, and SCFA can play a key role in generating the needed data.

Finding. Technologies for treating contaminant mixtures are in short supply, and the efficacy of many technologies for treating mixtures has not been established. The permeable reactive barrier is the most promising method for preventing the migration of mixtures of dissolved metals, radionuclides, and DNAPL components under appropriate conditions, but the longevity of barrier materials needs to be established.

- *Recommendation:* SCFA should fund tests designed to develop and determine performance limits for technologies capable of treating the types of contaminant mixtures that occur at DOE sites. In particular, SCFA should continue to fund studies of the longevity of reactive barriers in terms of reactivity, permeability, and integrity.

Finding. Removing all sources of groundwater contamination, particularly DNAPLs, will be technically impracticable at a number of DOE sites, and long-term containment systems will be necesaary for these sites. Methods will be needed to monitor the performance of containment barriers, because the longevity of barrier materials is uncertain. Electrical resistance tomography methods have received a significant amount of SCFA funding for studies of the integrity of subsurface barriers (including reactive barriers and conventional containment systems), but these methods have considerable limitations for this application.

- ***Recommendation:*** SCFA should focus a portion of the program's work on development of remedial alternatives (including containment systems) that prevent migration of contaminants at sites where contaminant source areas cannot be treated. Methods for monitoring long-term performance of these systems should be included in this work. In making its funding decisions, SCFA should distinguish between characterization technologies that can evaluate subsurface barrier performance and those that can delineate site features that are important in remedy selection and design.

Finding. Use of monitored natural attenuation in place of active cleanup remedies is increasing at contaminated sites nationwide, but implementing natural attenuation at DOE sites to help control plume migration may require additional research to develop methods for predicting the fate of certain classes of contaminants in natural environmental media.

- ***Recommendation:*** SCFA should determine what additional research will be needed for DOE to consider use of monitored natural attenuation at some of its sites, while still meeting applicable regulatory requirements, and should develop a corresponding research strategy.

Finding. Representative successful SCFA achievements in developing technologies for remediation of metals and radionuclides include in situ redox manipulation for chromium immobilization at Hanford, bottom barriers for waste containment at the Idaho National Engineering and Environmental Laboratory, the site characterization and analysis penetrometer system for characterization of subsurface environments in a wide range of settings, and the dig-face system for real-time guidance of excavations. Representative successful SCFA projects for developing DNAPL remediation technologies include dynamic underground stripping and thermally enhanced vapor extraction. Unfortunately, in the past SCFA has had to discontinue funding for some promising projects before technology development work was completed.

- ***Recommendation:*** SCFA should emphasize moving these and other technologies with demonstrated performance records to wider use. For its most promising projects, SCFA should ensure funding consistent with needs, including time for completion and long-term monitoring of field tests.

IMPROVING OVERALL PROGRAM DIRECTION

Finding. The overall goals of SCFA's technology development program have to be better defined in order to evaluate success. SCFA has struggled to provide Congress and others with concrete measures of program performance. To date, SCFA has focused on demonstrating the extent to which its technologies have been deployed in the field, but the available deployment data are inadequate (see Chapter 5). Further, the total number of SCFA technology deployments is not a sufficient metric for evaluating the SCFA program.

- ***Recommendation 1:*** SCFA should continue its efforts to work more closely with end users of remediation technologies (the DOE field personnel responsible for selecting these technologies) in setting its overall program direction. Working with end users, SCFA should identify key technical gaps and prepare a national plan for developing technologies to fill these gaps. This plan should be updated periodically as regulatory requirements and technology needs change. The extent to which the technology gaps have been filled should serve as the key measure of SCFA's success. SCFA's recent work with the site technology coordination groups to develop lists of technology needs (see Chapter 5) represents an important step in this direction, but implementation of the process has been hampered by budget limitations. Although SCFA developed a prioritized list of problem areas (known as work packages) for funding in fiscal year 1998, it was unable to use this list to guide its program because the entire SCFA budget went to supporting multiyear projects that began before SCFA was formed.
- ***Recommendation 2:*** SCFA should significantly increase use of peer review for (1) determining technology needs and (2) evaluating projects proposed for funding (see NRC, 1998, for guidelines on peer review). Peer reviews should carry sufficient weight to affect program funding.
- ***Recommendation 3:*** SCFA should improve the accuracy of its reporting of technology deployments. SCFA should use a consistent definition of deployment and should work with the Office of Environmental Restoration to verify the accuracy of its deployment report.

Finding. SCFA technology development projects have been most successful when they have been based on specific needs identified by

DOE installations and have involved DOE end users in planning the demonstrations. Other factors important in successful projects include (1) availability of sufficient financial support for timely demonstrations of the technology; (2) conduct of exploratory and pilot-scale assessments to enhance the technology prior to full-scale demonstrations; (3) adequacy of system design and operation; (4) availability of data showing cost savings; (5) multiplicity of potential applications; and (6) use of independent peer review in planning and evaluating demonstrations.

- ***Recommendation 1:*** SCFA should strive to increase the involvement of technology end users in planning the technology demonstrations it funds. End users should be involved in planning every demonstration that SCFA funds, as in the Accelerated Site Technology Deployment Program.
- ***Recommendation 2:*** SCFA should continue efforts to improve its success metrics for individual technology development projects. The metrics should be based on a careful analysis of factors that have led to the success or failure of past projects and could include the factors listed in the finding above.
- ***Recommendation 3:*** SCFA should identify successful technology demonstration projects to serve as models for future demonstrations.

Finding. Regulatory policies concerning cleanup requirements for groundwater and soil are evolving rapidly toward more flexible approaches. These policies will affect the range of cleanup goals that are acceptable at DOE installations and, correspondingly, the suite of possible remediation technologies for achieving these goals.

- ***Recommendation:*** SCFA should work with DOE field personnel to develop a process for continuously tracking cleanup requirements (and corresponding technology needs) at the sites. SCFA should keep track of policy changes, as discussed in Chapter 2 of this report, that may affect cleanup requirements.

OVERCOMING BARRIERS TO DEPLOYMENT

Finding. Contractors at DOE installations are reluctant to try innovative technologies developed by SCFA and others in part because of uncertainties about technology performance and the risk that the innovative technology will fail to perform as predicted.

- ***Recommendation 1:*** SCFA should sponsor more field demonstrations, such as those funded under the Accelerated Site Technology Deployment Program, to obtain credible performance and cost data. SCFA should consider whether sponsorship could include partial reimbursement for failed demonstrations, if an alternate remediation system has to be constructed to replace the failed one.
- ***Recommendation 2:*** SCFA should ensure that the project reports it provides contain enough technical information to evaluate potential technology performance and effectiveness relative to other technologies. The project descriptions contained in SCFA's periodic technology summary reports are not sufficiently detailed to serve this purpose. Project reports should include well-documented performance data, detailed cost estimates, design information useful to practitioners, and lessons learned. They should follow the guidelines in the Federal Remediation Technologies Roundtable's *Guide to Documenting and Managing Cost and Performance Information for Remediation Projects* (EPA, 1998).
- ***Recommendation 3:*** A key future role for the SCFA should be the development of design manuals for technologies that could be widely used across the weapons complex. SCFA could use the Air Force Center for Environmental Excellence (AFCEE) design manual for bioventing as a model (Leeson and Hinchee, 1996). The AFCEE approach is to test a technology at a number of well-characterized sites and develop design manuals from the results. Other possible models include the WASTECH® monograph series published by the American Academy of Environmental Engineers (AAEE)[1] and the surfactant-cosolvent manual (Lowe et al., 1999) published by the Advanced Applied Technology Demonstration Facility (AATDF, sponsored by the Department of Defense and based at Rice University).
- ***Recommendation 4:*** Appropriately qualified SCFA staff members (with in-depth knowledge of remediation technologies) should be available to serve as consultants on innovative technologies for DOE's environmental restoration program. These staff members also should develop periodic advisories for project managers on new, widely applicable technologies. SCFA needs to ensure that DOE technology end users are provided with

---

[1] Information about these monographs is available from the AAEE in Annapolis, Maryland.

the technical assistance (from within and as necessary from outside DOE) required to deploy new technologies at full scale.

Finding. Fewer than one-third of SCFA technologies have been deployed at more than one facility, and fewer than 20 percent have been deployed at more than two facilities. This lack of multisite deployments is primarily a result of the lack of demand for SCFA technologies, but lack of organized data on the types and locations of different contamination problems in the weapons complex also hinders multisite deployments. Without a well-organized data base on the prevalence of different types of subsurface contamination problems in the weapons complex, planning for multisite deployments is difficult.

- *Recommendation:* SCFA should use the potential for multisite application of new technologies as an important criterion in selecting projects for funding, although single applications are appropriate for unique, high-risk situations. DOE could strengthen its efforts to organize site characterization data so they can be easily accessed and used in planning SCFA's program.

ADDRESSING BUDGET LIMITATIONS

Finding. SCFA's progress has been limited in part by large budget swings. In fiscal year 1998, SCFA's budget was reduced to a level that was insufficient to support significant progress on the development of innovative remediation technologies. The budget was cut from a 1994 level of $82 million to a 1998 level of $15 million, which included a $5 million congressional earmark, leaving an effective budget of $10 million. This budget was inadequate to fund the types of large-scale demonstrations needed to transition innovative remediation technologies from the research and development phase to full-scale application. It also was too small to allow open bidding for project funding. The fiscal year 1999 budget of $25 million, while representing a significant increase, will allow for funding of only a limited number of projects.

- *Recommendation 1:* DOE managers should reassess the priority of subsurface cleanup relative to other problems and, if the risk is sufficiently high, should increase remediation technology development funding accordingly.
- *Recommendation 2:* SCFA should pursue a variety of strategies to leverage its funding. First, it should develop an im-

proved strategy for collaborating with external technology developers to adapt technologies for DOE use that have been developed in the private sector or by other government agencies. SCFA should work closely with the original technology developers to avoid duplicating their work, and new technology development efforts should focus on problems for which no cost-effective technical solutions exist. Second, SCFA should create stronger ties with the Environmental Management Science Program (EMSP). SCFA should assess the relevance of EMSP research for application to the SCFA program. Third, SCFA should continue its participation with working groups of the Remediation Technologies Development Forum (RTDF), a public-private partnership organization involved in remediation technology development. SCFA is involved in several RTDF working groups, including the Lasagna Partnership and the Permeable Reactive Barriers Action Team, and this involvement should continue.

In summary, DOE faces the challenge of cleaning up large quantities of contaminated groundwater and soil with a suite of baseline technologies that are not adequate for the job. Political pressure to meet federal and state groundwater and soil remediation requirements at DOE installations continues and recently has created problems for DOE at facilities such as Hanford, where politicians have pressured the department for better efforts to clean up contamination in the vadose zone (the soil above the water table). Although the implementation of site remediation laws is becoming somewhat more flexible, addressing groundwater and soil contamination problems at DOE installations cannot be avoided. DOE will have to continue to invest in accessing and developing remediation technologies for these media.

A number of new remediation technologies are currently in the pipeline that, with adequate DOE investment to complete development work, could make significant contributions to the cleanup effort. SCFA has overseen some successful technology development projects in the past. Although its operations need continued improvements as discussed in this report, nonetheless SCFA is the key entity within DOE for ensuring that the department will be adequately equipped to solve its groundwater and soil contamination problems. DOE managers as a whole need to reassess the priority assigned to subsurface remediation technology development and whether SCFA is adequately supported and organized to accomplish its mission.

REFERENCES

EPA (Environmental Protection Agency). 1998. Guide to Documenting and Managing Cost and Performance Information for Remediation Projects. EPA 542-B-98-007. Washington, D.C.: EPA.

Leeson, A., and R. E. Hinchee. 1996. Principles and Practices of Bioventing. Brooks Air Force Base, Tex.: U.S. Air Force Center for Environmental Excellence.

Lowe, D. F., C. L. Oubre, and C. H. Ward, Eds. 1999. Surfactants and Cosolvents for NAPL Remediation: A Technology Practices Manual. Boca Raton, Fla.: Lewis Publishers.

NRC (National Research Council). 1998. Peer Review in Environmental Technology Development Programs. The Department of Energy's Office of Science and Technology. Washington, D.C.: National Academy Press.

Appendixes

A

Facilities Where DOE is Responsible for Environmental Cleanup

State	Name	Location	Weapons Production Activities
AK	Amchitka Island	Amchitka	This area served as an underground nuclear weapons testing site for three test shots in 1965, 1969, and 1971.
AZ	Monument Valley	Monument Valley	Between 1955 and 1967, a uranium mill at this site produced an upgraded uranium product that was further milled at a uranium mill in Shiprock, New Mexico, eventually producing uranium concentrate for sale to the Atomic Energy Commission (AEC).
AZ	Tuba City	Tuba City	Between 1955 and 1966, a uranium mill at this facility processed uranium concentrate for sale to AEC.
CA	Lawrence Livermore National Laboratory (LLNL)–Main Site	Livermore	LLNL is composed of two sites, the Main Site and Site 300. The Main Site, initially used as a flight training base and engine overhaul facility, began to be used for nuclear weapons research in 1950.
CA	Lawrence Livermore National Laboratory–Site 300	Livermore	This site is used as a remote high-explosives testing area. It includes several areas for high-explosive component testing, several instrument firing tables, a particle accelerator, and various support and service facilities.
CA	Oxnard Site	Oxnard	A Department of Energy (DOE) contractor occupied the site between 1981 and 1984 to produce forgings for weapons parts. DOE purchased the site in 1984 and continued to produce forgings until 1995.
CA	Sandia National Laboratories, California	Alameda County	This site was established by AEC in 1956 to conduct research and development in the interest of national security with emphasis on nuclear weapons development and engineering

			in cooperation with Lawrence Livermore National Laboratory.
CO	Durango	Durango	Initially the site of a vanadium production plant, this site milled uranium ore for Manhattan Engineer District (MED) and AEC between 1943 and 1963.
CO	Grand Junction Mill Tailings Site	Grand Junction	Between 1951 and 1967, a uranium mill at this site processed uranium ore, producing uranium concentrate for sale to AEC. The site also produced vanadium and milled uranium for commercial sale until 1970.
CO	Grand Junction Projects Office	Grand Junction	MED established this site in 1943 to refine uranium for the federal government. Between 1947 and 1970, the site administered AEC defense-related uranium exploration and purchase programs.
CO	Gunnison	Gunnison	Between 1958 and 1962, a uranium mill at this site processed uranium ore, producing uranium concentrate for sale to AEC.
CO	Maybell	25 miles west of Craig	Between 1955 and 1964, a uranium mill at this site processed uranium ore, producing uranium concentrate for sale to AEC.
CO	Naturita	Naturita	Between 1947 and 1958, a uranium mill at this site processed uranium ore, producing uranium concentrate for sale to AEC. Between 1961 and 1963, the site produced a uranium product that was further processed at a uranium mill in Durango, Colorado, eventually producing uranium concentrate for sale to AEC.

State	Name	Location	Weapons Production Activities
CO	Old and New Rifle	Rifle	Between 1948 and 1970, two uranium mills at these sites processed uranium ore, producing uranium concentrate for sale to AEC.
CO	Rocky Flats Environmental Technology Site	16 miles northwest of Denver	Established in 1952 as the Rocky Flats Plant, this site produced the plutonium pits used as triggers in nuclear weapons as well as other uranium, beryllium, and steel weapons components. Rocky Flats also recovered plutonium from returned weapons parts, production scrap, and residues.
CO	Slick Rock	Slick Rock	Two uranium mills operated at this site. The first, which operated between 1931 and 1943, was a vanadium and radium mill that also produced uranium for MED. Between 1957 and 1961, a second uranium mill nearby processed uranium ore, producing a uranium product that was further milled at one of the uranium mills at Rifle, Colorado, eventually producing uranium concentrate for sale to AEC.
FL	Pinellas Plant	St. Petersburg	Between 1957 and 1994, this site produced precisely timed neutron generators to initiate nuclear devices and other nonnuclear weapons parts.
HI	Kauai Test Facility	Kauai	Sandia National Laboratory/New Mexico has conducted some nonnuclear weapons research and development at this site, including launching rockets carrying experimental non-nuclear payloads.

IA	Ames Laboratory	Ames	Located on the campus of Iowa State University, this site developed and operated the first efficient production-scale process to convert uranium tetrafluoride to metal for use as reactor fuel by MED.
ID	Idaho National Engineering and Environmental Laboratory (INEEL)	Approximately 42 miles northwest of Idaho Falls	AEC established the National Reactor Testing Station in 1949 on the site of a 1940s U.S. Navy bombing and artillery range. Today, the site is known as the Idaho National Engineering and Environmental Laboratory. Between 1953 and 1992, the Idaho Chemical Processing Plant at INEEL reprocessed spent fuel from naval propulsion, test, and research reactors to recover enriched uranium for reuse in nuclear weapons production. Large volumes of transuranic and low-level waste from Rocky Flats Plants (RFP) component fabrication operations are buried and stored at INEEL, including waste resulting from two fires at RFP. Facilities at INEEL also conducted various minor nuclear weapons research and development work.
ID	Lowman	Lowman	Between 1956 and 1960, a uranium mill at this site processed mineral processing residues, producing uranium for sale to AEC. The source of contamination was residual tailings. The site also produced other specialty minerals for weapons and nonweapons use.
KY	Paducah Gaseous Diffusion Plant	Paducah	Built in the early 1950s, this plant was initially operated for the sole purpose of enriching uranium for weapons production. It gradually began to supply enriched uranium for Navy and commercial reactor fuel as well. Until the early 1960s, UF_6 feed for the diffusion process was also produced at the site. In accordance with the Energy Policy Act of 1992, the diffusion cascade and support facilities at

State	Name	Location	Weapons Production Activities
			the site have been leased to the government-owned United States Enrichment Corporation since 1993. Paducah is still in operation enriching uranium for commercial customers, primarily nuclear power utilities.
MO	Kansas City Plant	Kansas City	Constructed in 1942 to build Navy aircraft engines, this site was converted to manufacture nonnuclear components for nuclear weapons in 1949. Today it continues to be DOE's main component fabrication plant.
MO	Weldon Spring Site Remedial Action Project	St. Charles County	Located on the site of a former ordnance production facility, this site operated from 1956 until 1966 to sample and refine uranium ore for AEC and to manufacture production reactor fuel.
MO	Belfield	Belfield	Between 1965 and 1967, a gas-fired rotary kiln at this site burned uraniferous lignite coal. The ash was shipped to a uranium mill in Rifle, Colorado, eventually producing uranium concentrate for sale to AEC.
NM	Ambrosia Lake	McKinley County	This facility was a uranium milling site built in 1957. It sold uranium to AEC between 1958 and 1969. Sources of contamination were residual tailings and discharged process water remaining after uranium was extracted during the milling process.
NM	Los Alamos National Laboratory	Los Alamos	Established in 1943 to design, develop, and test nuclear weapons, Los Alamos also produced small quantities of plutonium metal and nuclear weapons components. Its focus now includes academic and industrial research.

NM	Sandia National Laboratories, New Mexico	Albuquerque	Established in 1949, this laboratory was formed from the Los Alamos Explosive Ordnance "Z Division" to design nonnuclear components of nuclear weapons. Sandia also housed a weapons assembly line from 1946 until 1957.
NM	Shiprock	Shiprock	Between 1954 and 1968, a uranium mill at this site processed uranium ore, producing uranium concentrate for sale to AEC.
NM	South Valley Site	Albuquerque	Between 1951 and 1967, this site, owned by AEC and known as South Albuquerque Works, fabricated nonnuclear components for nuclear weapons. The site was later transferred to the Air Force for use as a jet engine factory and eventually sold to General Electric.
NV	Central Nevada Test Site	60 miles northeast of Tonopah	This site was used for one subsurface nuclear test and nonnuclear seismic experiments.
NV	Nevada Test Site	65 miles northwest of Las Vegas	Established in 1950, the Nevada Test Site was used for full-scale atmospheric and underground testing of nuclear explosives in connection with weapons research and development. It is also currently used for disposal of low-level radioactive waste from DOE sites.
NV	Tonopah Test Range	Nellis Air Force Range	This site assumed the function of the Salton Sea Test Base in 1961. It is used by Sandia National Laboratories, New Mexico, to test the mechanical operation and delivery systems for nuclear weapons and other defense-related projects.
OH	Fernald Environmental Management Project (FEMP)	Fernald	FEMP was established as the Feed Materials Production Center in the early 1950s to convert uranium ore into uranium metal and to fabricate uranium metal into target

State	Name	Location	Weapons Production Activities
			elements for reactors that produced plutonium and tritium. The site ceased production in 1989.
OH	Mound Plant	Miamisburg	Beginning in 1946, this government-owned site developed and fabricated nuclear and nonnuclear components for the weapons program, including polonium-beryllium initiators. In the 1950s, the plant began to build detonators, cable assemblies, and other nonnuclear products. Mound began to retrieve and recycle tritium from dismantled nuclear weapons in 1969. Nonweapons activities included the production of plutonium-238 thermoelectric generators for spacecraft.
OH	Portsmouth Gaseous Diffusion Plant	Portsmouth	Built in the early 1950s, this site initially produced highly enriched uranium for weapons. Later, the high-enrichment portion of the diffusion cascade was used to produce highly enriched uranium (HEU) for naval propulsion and research and test reactors, and eventually shut down. In accordance with the Energy Policy Act of 1992, the lower portion of the diffusion cascade and support facilities at the site have been leased to the government-owned United States Enrichment Corporation. These facilities are still in operation enriching uranium for commercial customers, primarily nuclear power utilities.
OH	RMI Titanium Company	Ashtabula	Between 1962 and 1988, this privately owned site received uranium billets from Fernald and extruded them into various shapes for reactor fuel and targets.
OR	Lakeview	Lakeview	Between 1958 and 1960, a uranium mill at this site processed uranium ore, producing uranium components for sale to

			AEC. In 1978, the mill was sold and used as a lumber mill and a stockpile area for sawdust and scrap waste.
PA	Canonsburg	Canonsburg	This site refined uranium for AEC.
SD	Savannah River Site (SRS)	Aiken	This site was established in 1950 to produce, purify, and process plutonium, tritium, and other radioisotopes for nuclear weapons programs and other purposes. The site fabricated fuel, operated five reactors and two chemical separation plants, and conducted research and development. SRS also produced heavy water and processed tritium. Nonweapons activities include production of plutonium-238 for use in thermoelectric generators.
SD	Edgemont Vicinity Properties	Edgemont	Between 1956 and 1968, a uranium mill at Edgemont milled uranium for AEC. The mill also produced vanadium and milled uranium for other customers until 1974. The mill site was cleaned up by the Tennessee Valley Authority and is not a DOE site, but DOE cleaned up vicinity properties under DOE's Uranium Mill Tailings Remediation Control Act (UMTRCA) program.
TN	K-25 Site	Oak Ridge	K-25 was built in 1943 and 1944 to supply enriched uranium for nuclear weapons production. It was later modified to produce commercial-grade low-enriched uranium. It has been shut down since 1987.
TN	Oak Ridge National Laboratory (ORNL)	Oak Ridge	In 1942, MED established research facilities in Oak Ridge to produce and separate the first gram quantities of plutonium. Since then, ORNL has primarily supported nonweapons programs, including radioisotope production and research in a variety of fields. ORNL has also supplied isotopes for the nuclear weapons program.

State	Name	Location	Weapons Production Activities
TN	Y-12 Plant	Oak Ridge	Originally established by MED to use an electromagnetic process to separate uranium isotopes, Y-12 later enriched lithium and fabricated and stored nuclear weapons components containing lithium and HEU.
TX	Falls City	46 miles southeast of San Antonio	Between 1961 and 1968, a uranium mill at the Falls City site milled uranium for AEC.
TX	Pantex Plant	Amarillo	Formerly a conventional munitions plant also used by Texas Tech University for nondefense activities, AEC converted this site to a high-explosives component fabrication and weapons assembly plant in 1951. The principal operation of Pantex is currently weapons disassembly and fissile material storage.
UT	Green River	Green River	Between 1958 and 1961, a uranium concentrator operating at this site produced an upgraded uranium product for subsequent milling at Rifle, Colorado, and eventual sale to AEC. The site also produced vanadium for nonweapons purposes.
UT	Mexican Hat	Mexican Hat	Between 1957 and 1965, a commercially owned uranium mill at this site processed uranium ore, producing uranium concentrate for sale to AEC.
UT	Monticello Site	Monticello	Between 1943 and 1960, a uranium mill at this site processed uranium ore, producing uranium concentrate for sale to AEC. The mill was commercially owned until 1948, when AEC purchased the facility.

UT	Salt Lake City	Salt Lake City	Between 1951 and 1964, a uranium mill at this site processed uranium ore, producing uranium concentrate for sale to AEC.
WA	Hanford	Richland	Established in 1942, this major government-owned nuclear weapons production site fabricated reactor fuel, operated nine reactors and five chemical separation facilities, and fabricated plutonium components for nuclear weapons. Later operations included nonmilitary applications of nuclear energy.
WY	Riverton	Riverton	Between 1962 and 1965, a uranium concentrator at this facility processed uranium ore, producing an upgraded uranium product that was further processed at Slide Rock, Colorado, eventually producing uranium concentrate for sale to AEC.
WY	Spook	Converse County	Between 1958 and 1963, a uranium mill at this facility processed uranium ore, producing uranium concentrate for sale to AEC.

SOURCE: Adapted from Linking Legacies: Connecting the Cold War Nuclear Weapons Processes to Their Environmental Consequences, DOE, Washington, D.C., 1997.

B

SCFA Technology Deployment Report

Project	Year	Site
Lasagna®	1997	Paducah
Six-Phase Soil Heating	1996	Chicago Commercial Site
Six-Phase Soil Heating	1996	Niagra Falls Municipal Airport
In Well Vapor Stripping	1997	Brookhaven National Laboratory
In Well Vapor Stripping	1997	Edwards Air Force Base
In Well Vapor Stripping	1997	Savannah River Site
Dynamic Underground Stripping	1995	Lawrence Livermore National Laboratory
Dynamic Underground Stripping	1995	Lawrence Livermore National Laboratory
Environmental Measurement While Drilling	1996	Savannah River Site
Dig Face Characterization	1996	Idaho National Engineering and Environmental Laboratory
Dig Face Characterization	1996	Idaho National Engineering and Environmental Laboratory
Dig Face Characterization	1995	Mound
Smart Geomembrane	1997	Idaho National Engineering and Environmental Laboratory
Passive Reactive Barrier	1998	Oak Ridge
Passive Reactive Barrier	1998	Rocky Flats
Thermal Enhanced Vapor Extraction System (TEVES)	1996	Idaho National Engineering and Environmental Laboratory
Thermal Enhanced Vapor Extraction System (TEVES)	1999	Lawrence Berkeley National Laboratory
Thermal Enhanced Vapor Extraction System (TEVES)	1999	Lawrence Berkeley National Laboratory
Thermal Enhanced Vapor Extraction System (TEVES)	1996	Sandia National Laboratories
Frozen Soil Barrier	1997	Oak Ridge
Frozen Soil Barrier	1997	Oak Ridge
Flameless Thermal Oxidation	ND	Idaho National Engineering and Environmental Laboratory
SEAMIST	ND	Hanford
SEAMIST	ND	Los Alamos National Laboratory
SEAMIST	1991	Lawrence Livermore National Laboratory
SEAMIST	1991	Lawrence Livermore National Laboratory
SEAMIST	1996	Savannah River Site
SEAMIST	ND	Yucca Mountain
Deep Soil Mixing	1997	Argonne National Lab
Deep Soil Mixing	1997	Argonne National Lab

Project	Year	Site
Deep Soil Mixing	1994	Portsmouth
Resonant Sonic Drilling	1994	Fernald Environmental Management Project
Resonant Sonic Drilling	1991	Hanford
Resonant Sonic Drilling	1992	Hanford
Resonant Sonic Drilling	1995	Hanford
Resonant Sonic Drilling	1995	Laboratory for Energy-Related Health Research
Resonant Sonic Drilling	1995	Laboratory for Energy-Related Health Research
Resonant Sonic Drilling	1995	Savannah River Site
Passive Soil Vapor Extraction (Barometric Pumping)	1995	Hanford
Passive Soil Vapor Extraction (Barometric Pumping)	1995	Idaho National Engineering and Environmental Laboratory
Passive Soil Vapor Extraction (Barometric Pumping)	1996	Savannah River Site
Passive Soil Vapor Extraction (Barometric Pumping)	1997	Savannah River Site
Electrokinetics—Arid	1996	Sandia National Laboratories
Electrokinetics—Arid	1996	Sandia National Laboratories
In Situ Vitrification Bottoms-up	1996	Oak Ridge
Automated Control System for Soil Vapor Extraction	1996	Savannah River Site
Recirculating Wells	1997	Porstmouth
Recirculating Wells	1997	Savannah River Site
In Situ Permeable Flow Sensor	1995	Edwards Air Force Base
In Situ Permeable Flow Sensor	1995	Hanford
In Situ Permeable Flow Sensor	1995	Savannah River Site
Cryogenic Cutting	1998	Hanford
Cryogenic Cutting	1996	Idaho National Engineering and Environmental Laboratory
HaloSnif	1995	Hanford
HaloSnif	1995	Tinker Air Force Base
Unsaturated Flow Apparatus	1994	Hanford
Surfactant/Alcohol Flushing for DNAPLs	1998	Lawrence Livermore National Laboratory
Surfactant/Alcohol Flushing for DNAPLs	1998	Lawrence Livermore National Laboratory
Surfactant/Alcohol Flushing for DNAPLs	1998	Lawrence Livermore National Laboratory
In Situ Anaerobic Bioremediation System	1996	Idaho National Engineering and Environmental Laboratory

In Situ Anaerobic Bioremediation System	1996	Idaho National Engineering and Environmental Laboratory
In Situ Anaerobic Bioremediation System	1996	Point Miguel
Cryogenic Drilling	1995	Lawrence Berkeley National Laboratory
Cryogenic Drilling	1995	Lawrence Berkeley National Laboratory
Cryogenic Drilling	1995	Lawrence Berkeley National Laboratory
Solution Mining	1997	Fernald
Smart Sampling	1996	Mound
In Situ Chemical Oxidation of Soils	1997	Argonne National Lab
In Situ Chemical Oxidation of Soils	1997	Argonne National Lab
In Situ Chemical Oxidation of Soils	1996	Kansas City Plant
In Situ Chemical Oxidation of Soils	1996	Kansas City Plant
Adsorption/Desorption Relative to Bioremediation of Chlorinated Solvents	1996	Kansas City Plant
Adsorption/Desorption Relative to Bioremediation of Chlorinated Solvents	1996	Portsmouth
Landfill Assessment and Monitoring System	1994	Sandia National Laboratories
Landfill Assessment and Monitoring System	1994	Sandia National Laboratories
Innovative DNAPL Characterization Technologies	1996	Savannah River Site
Biomass Remediation System	1997	Formerly Utilities/Sites Remedial Action Program
Colloidal Borescope	1994	Lawrence Livermore National Laboratory
Colloidal Borescope	1994	Lawrence Livermore National Laboratory
Colloidal Borescope	1992	Oak Ridge
Fiber Optic Probe for Trichloroethylene in Soil and Groundwater	1993	Hanford
Fiber Optic Probe for Trichloroethylene in Soil and Groundwater	1992	Lawrence Livermore National Laboratory
Fiber Optic Probe for Trichloroethylene in Soil and Groundwater	1992	Lawrence Livermore National Laboratory
Biomolecular Probe Analysis: Bioremediation Organisms	1995	Savannah River Site
Destruction of Carbon Tetrachloride by Steam Reforming	1992	Oak Ridge

Project	Year	Site
Destruction of Carbon Tetrachloride by Steam Reforming	1995	Oak Ridge
Heavy Weight Cone Penetrometer	ND	Argonne National Lab
Heavy Weight Cone Penetrometer	1993	Hanford
Heavy Weight Cone Penetrometer	1993	Hanford
Heavy Weight Cone Penetrometer	1993	Hanford
Heavy Weight Cone Penetrometer	1993	Hanford
Heavy Weight Cone Penetrometer	1993	Hanford
Heavy Weight Cone Penetrometer	1993	Hanford
Heavy Weight Cone Penetrometer	1993	Hanford
Heavy Weight Cone Penetrometer	1993	Hanford
Heavy Weight Cone Penetrometer	1995	Pantex
Rapid Transuranic Monitoring Laboratory	1995	Formerly Utilities/Sites Remedial Action Program
Rapid Transuranic Monitoring Laboratory	1995	Formerly Utilities/Sites Remedial Action Program
Rapid Transuranic Monitoring Laboratory	1993	Idaho National Engineering and Environmental Laboratory
Arid Site Drilling Technology Development	1992	Hanford
Advanced In Situ Moisture Logging System	1994	Lawrence Livermore National Laboratory
Advanced In Situ Moisture Logging System	1994	Lawrence Livermore National Laboratory
Advanced In Situ Moisture Logging System	1993	Sandia National Laboratories
Cross Borehole Electromagnetic Imaging	1995	Sandia National Laboratories

NOTE: DNAPL = dense-nonaqueous phase liquid; ND = technology was slated for deployment but had not yet been deployed as of January 14, 1998, when this listing was prepared.

SOURCE: Department of Energy Subsurface Contaminants Focus Area, unpublished data, January 14, 1998.

C

Biographical Sketches of Committee Members and Staff

COMMITTEE MEMBERS

C. HERB WARD, who chaired the committee, is the Foyt family chair of engineering at Rice University, where he is also professor of environmental science and engineering and ecology and evolutionary biology. He directs the Energy and Environmental Systems Institute, the Department of Defense Advanced Applied Technology Demonstration Facility, and the National Center for Ground Water Research. In addition, he serves as codirector of the U.S. Environmental Protection Agency (EPA)-sponsored Hazardous Substances Research Center/South and Southwest. His research interests include the microbial ecology of hazardous waste sites, biodegradation by natural microbial populations, microbial processes for aquifer restoration, and microbial transport and fate. He also chairs the National Research Council's Committee on Peer Review in the Department of Energy—Office of Science and Technology. He received his Ph.D. in plant pathology, genetics, and physiology from Cornell University and an M.P.H. in environmental health from the University of Texas.

HERBERT E. ALLEN is a professor in the Department of Civil and Environmental Engineering and a professor of oceanography in the Graduate College of Marine Studies at the University of Delaware. His research interests include environmental chemistry, fate and effects of pollutants in water, sediment and soil environments, development of environmental standards, analytical chemistry, and the hazardous treatment of explosives and metals. He is a visiting professor in the Department of Environmental Science at Nankai University in the People's Republic of China and had a World Health Organization fellowship to study environmental chemistry in The Netherlands and Germany. He received a B.S. in chemistry from the University of Michigan in 1962, an M.S. in analytical chemistry from Wayne State University in 1967, and a Ph.D. in environmental chemistry from the University of Michigan in 1974.

RICHARD E. BELSEY is an emeritus professor of pathology at the Oregon Health Sciences University. His medical training is in internal medicine and endocrinology, and his research interests focus on the health and safety issues associated with activities at the Hanford Site. He is a member of the Portland Chapter of Physicians for Social Responsibility and a member of its National Task Force on Nuclear Weapons and Public Health. He has been a Hanford Advisory Board member since 1994 and chair of its Health, Safety and Waste Management Committee. He has served on the State of Oregon Hanford Waste Board since 1990 and is chair of its Waste Cleanup and Site Restoration Committee. He received his M.D. from the Albany Medical College in 1966.

KIRK W. BROWN is a professor of soil science at Texas A&M University and is also a member of the faculty of toxicology. He is a consultant with K. W. Brown Environmental Services, which he founded in 1981. His research focuses on the land disposal of wastes and the cleanup of sites contaminated with agricultural and industrial chemicals. He has served on several EPA, Office of Technology Assessment, and National Research Council committees and has received numerous awards from Texas A&M and from professional societies. He received his B.S. from Delaware Valley College, his M.S. from Cornell University, and his Ph.D. from the University of Nebraska.

RANDALL J. CHARBENEAU is a professor in the Department of Civil Engineering at the University of Texas and director of the Center for Research on Water Resources. His expertise is in groundwater pollution, fate and transport, and modeling, and his research interests include groundwater hydraulics and contaminant transport, numerical modeling, and radiological assessment. He has served on numerous panels, including review of an incineration risk assessment at the Savannah River Site, a review of performance evaluation of mixed low-level waste disposal sites, and a project review of in situ redox manipulation. He received a B.S. from the University of Michigan in

1973, an M.S. from Oregon State University in 1975, and a Ph.D. from Stanford University in 1978, all in civil engineering.

RICHARD A. CONWAY is an environmental consultant and retired senior corporate fellow at Union Carbide Corporation. His areas of expertise include contaminated site remediation, hazardous waste management, and environmental risk analysis of chemical products. He was elected to the National Academy of Engineering in 1986 for his contributions to environmental engineering and for the development of improved treatment processes for industrial wastes. He has received many awards and honors, including the Hering Medal, Gascoigne Medal, Dudley Medal, Rudolfs Medal, and honors from the American Society of Civil Engineers, the Water Environment Federation, and the American Society for Testing and Materials. He has been involved in numerous NRC activities, including the Board on Environmental Studies and Toxicology, the Water Science and Technology Board, and the Committee on Peer Review in the Department of Energy—Office of Science and Technology. He received his M.S. in environmental engineering from the Massachusetts Institute of Technology.

HELEN E. DAWSON is an assistant professor in the Department of Environmental Science and Engineering at the Colorado School of Mines. Her research interests include transport and fate of organic and inorganic contaminants, solute transport in saturated and unsaturated sediments, and transport and remediation of petroleum hydrocarbons and chlorinated solvents as free phases in subsurface systems. She received a B.S. in geology from Stanford University in 1987, an M.S. in geochemistry from the Colorado School of Mines, and a Ph.D. in environmental engineering from Stanford University.

JOHN C. FOUNTAIN is a professor of geochemistry at the State University of New York at Buffalo. His research focuses on various aspects of contaminant hydrology, including aquifer remediation and the characterization of fractured rock aquifers. He is also a member of the NRC's Committee on Peer Review in the Department of Energy—Office of Science and Technology. He received his B.S. from California Polytech State University, San Luis Obispo, in 1970, his M.A. in 1973, and his Ph.D. in geology in 1975, both from the University of California at Santa Barbara.

RICHARD L. JOHNSON is an associate professor in the Department of Environmental Science and Engineering at the Oregon Graduate Institute and directs the Center for Groundwater Research. He researches the processes that control the movement of subsurface contaminants in the environment. He received a B.S. in chemistry from the University of Seattle in 1973, an M.S. in 1981 and a Ph.D. in 1984, both in environmental science from the Oregon Graduate Center.

ROBERT D. NORRIS is the technical director of bioremediation services at Eckenfelder, Brown and Caldwell in Nashville, Tennessee. He has managed numerous remediation projects and served as a technical expert on many projects for both EPA- and state-mandated remedial actions, feasibility studies, and treatability studies for a wide range of in situ and ex situ remediation technologies. Currently, he is managing the implementation of a zero-valence metal-permeable barrier at a Resource Conservation and Recovery Act site for treatment of chlorinated volatile organic compounds and chromium, and evaluation of this technology at a Department of Energy site for treatment of trichloroethylene, uranium, and technetium. He holds 13 patents, 4 on various aspects of bioremediation. He received a B.S. in chemistry from Beloit College in 1966 and a Ph.D. in chemistry from the University of Notre Dame in 1971.

FREDERICK G. POHLAND is professor and Edward R. Weidlein chair of environmental engineering at the University of Pittsburgh. His research has focused on environmental engineering operations and processes, solid and hazardous waste management, and environmental impact monitoring and assessment. He was elected to the National Academy of Engineering in 1993 for advancing the theory of anaerobic treatment processes and applications to solid waste management. He has been a visiting scholar at the University of Michigan and a guest professor at the Delft University of Technology in The Netherlands. He received his B.S. in civil engineering from Valparaiso University and his M.S. and Ph.D. in environmental engineering from Purdue University.

KARL K. TUREKIAN is Benjamin Silliman professor in the Department of Geology and Geophysics at Yale University and has directed the Center for the Study of Global Change. His expertise is in the geochemistry of radionuclides and trace elements and marine geochemistry. He was elected to the National Academy of Sciences in 1984. He has participated in many NRC activities, including the Commission on Geosciences, Environment, and Resources; the Commission on Physical Sciences, Mathematics, and Resources; the U.S. Committee for Geochemistry; and the Ocean Studies Board. He received his A.B. from Wheaton College in 1949, his M.A. in 1951, and his Ph.D. in 1955, both from Columbia University.

JOHN C. WESTALL is a professor in the Department of Chemistry at Oregon State University and an adjunct professor in the Department of Science and Engineering at the Oregon Graduate Institute. His research focuses on the application of surface and solution chemistry to problems in environmental chemistry, electrochemistry, and analytical chemistry, particularly the complex interactions of metals with organic materials and soil and the development of models for these interactions. He received a B.S. in chemistry from the University of North Carolina in 1971 and a Ph.D. in chemistry from the Massachusetts Institute of Technology in 1977.

STAFF

JACQUELINE A. MACDONALD is associate director of the National Research Council's Water Science and Technology Board. She directed the studies that led to the reports *Innovations in Ground Water and Soil Cleanup: From Concept to Commercialization, Alternatives for Ground Water Cleanup, In Situ Bioremediation: When Does It Work?, Safe Water from Every Tap: Improving Water Service to Small Communities,* and *Freshwater Ecosystems: Revitalizing Educational Programs in Limnology.* She received the 1996 National Research Council Award for Distinguished Service. Ms. MacDonald earned an M.S. degree in environmental science in civil engineering from the University of Illinois, where she received a university graduate fellowship and Avery Brundage scholarship, and a B.A. degree magna cum laude in mathematics from Bryn Mawr College. She has written about environmental remediation technologies for a number of publications, including *Environmental Science and Technology, Water Environment and Technology,* and *Soil and Groundwater Cleanup.*

D

Acronyms

AAEE	American Academy of Environmental Engineers
AATDF	Advanced Applied Technology Demonstration Facility
ACL	Alternate concentration limit
AFCEE	Air Force Center for Environmental Excellence
ALE	Arid Lands Ecology
ARAR	Applicable or relevant and appropriate requirement
ASTM	American Society for Testing and Materials
BMPS	Best management practices
BWCS	Buried waste containment system
CBO	Congressional Budget Office
CEMT	Committee on Environmental Management Technologies
CERCLA	Comprehensive Environmental Response, Compensation, and Liability Act (1980)
CERE	Consortium for Environmental Risk Evaluation
CFC	Chlorofluorocarbon
CMI	Corrective measures implementation
CMS	Corrective measures study
CRADA	Cooperative research and development agreement
CWA	Clean Water Act (1974)
DART	Decision Analysis for Remediation Technologies
DCA	Dichloroethane
DCE	Dichloroethylene
DNAPL	Dense nonaqueous-phase liquid
DOD	Department of Defense
DOE	Department of Energy

DSM	Deep-soil mixing
DTPA	Diethylenetriaminepentaacectic acid
DUS	Dynamic underground stripping
EDTA	Ethylenediaminetetraacetic acid
EMSP	Environmental Management Science Program
EPA	Environmental Protection Agency
GAO	General Accounting Office
GPR	Ground penetrating radar
HEU	Highly enriched uranium
IAEA	International Atomic Energy Agency
INEEL	Idaho National Engineering and Environmental Laboratory
ISRM	In situ redox manipulation
ISV	In situ vitrification
ITMS	Ion trap mass spectrometry
LLNL	Lawrence Livermore National Laboratory
LNAPL	Light nonaqueous-phase liquid
MCL	Maximum contaminant level
NAPL	Nonaqueous-phase liquid
NCP	National Contingency Plan
NOM	Natural organic matter
NPL	National Priorities List
NRC	National Research Council
O&M	Operations and maintenance
OST	Office of Science and Technology
OTA	Office of Technology Assessment
PA/SI	Preliminary assessment/site inspection
PAH	Polycyclic aromatic hydrocarbon
PCB	Polychlorinated biphenyl
PCE	Perchloroethylene
PNNL	Pacific Northwest National Laboratory
RAHCO	R.A. Hanson Company
RAPIC	Remediation Action Program Information Center
RBCA	Risk-based corrective action
RBSL	Risk-based screening level
RCRA	Resource Conservation and Recovery Act (1976)
RD/RA	Remedial design/remedial action
RFA	RCRA facility assessment
RFI	RCRA facility investigation
RI/FS	Remedial investigation/feasibility study
ROD	Record of decision
ROI	Radius of influence
RS	Remedial selection

RTDF	Remediation Technologies Development Forum
SARA	Superfund Amendment and Reauthorization Act (1986)
SCAPS	Site characterization and analysis penetrometer system
SCFA	Subsurface Contaminants Focus Area
SDWA	Safe Drinking Water Act (1974)
SITE	Superfund Innovative Technology Evaluation
SSL	Soil screening level
SSTL	Site-specific target level
STCG	Site technology coordination group
SVE	Soil vapor extraction
SVOC	Semivolatile organic compound
TCA	Trichloroethane
TCE	Trichloroethylene
TCLP	Toxicity characteristic leaching procedure
TEVES	Thermally enhanced vapor extraction system
UMTRCA	Uranium Mill Tailings Remediation Control Act (1978)
VC	Vinyl chloride
VOC	Volatile organic compound

Index

A

Accelerated Site Technology Deployment Program, 12, 212, 213-214, 218, 246
Accelerating Cleanup: Paths to Closure, 36
Advanced Applied Technology Demonstration Facility, 13
Air Force Center for Environmental Excellence, 60, 190, 246
Air pollution, incineration of contaminants, 19
Air sparging, 7, 135, 147-150, 215, 242
Alcohol flushing, 150-154, 266
Alternate concentration limits, 48-49
American Academy of Environmental Engineers, 13, 246
American Society for Testing and Materials, 65, 66
American Society of Mechanical Engineers, 209
Applicable or relevant and appropriate requirement, 48, 49, 51
Army Corps of Engineers, 211, 220
Army Environmental Center, 227
Atoms for Peace Program, 25
Attenuation, *see* Natural attenuation

B

Barrier technologies, *see* Subsurface barriers
Bioremediation, 215, 216, 217, 266, 267
  DNAPLs, 8, 136, 173-180, 185, 187, 189, 194
  metals and radionuclides, 6, 78, 110-112, 121
  see also Natural attenuation
Bioventing, 13, 140-141, 246
Brownfield sites, 4, 54, 63-64, 69
Budgetary issues, *see* Funding
Buried waste containment system, 225, 226

C

Canada, 161
Carbon tetrachloride, 29, 156, 184
Chlorinated solvents, *vii*, 4, 7, 8, 9, 30, 129, 130, 133, 149, 158, 168, 173, 176-178, 182, 183, 184, 186, 189-191, 193
  see also specific substances
Chlorofluorocarbons, 29
Chromatography, 82
Chromium, 5, 27, 29, 33, 48, 50, 73, 74, 76, 106, 110, 111, 222-225, 243
Clean Water Act, 42

Climatic conditions, 27, 181
Comprehensive Environmental Response, Compensation, and Liability Act, 3, 10, 40, 41, 43-51, 53, 54, 61, 62, 68, 69, 203, 206
  brownfields programs, 63, 64
  monitored natural attenuation policy, 58, 189, 191
  records of decision, 46, 52, 53, 54, 56, 61-63, 69
  risk-based assessment policies, 67
  technical impracticability waivers, 55, 56, 57
Computer applications, 80-82
Consortium for Environmental Risk Evaluation, 30-31, 32-33, 34-35
Containment systems, *see* Subsurface barriers
Contractors and contracting, 9, 10, 15-16, 204, 205, 207, 211, 218-219, 225, 236, 241, 245
Cooperative research and development agreements, 10, 225, 227, 236
Cosolvent flushing, 8, 150-154, 194, 241
Cost and cost-benefit factors
  CERCLA, 46
  Department of Energy spending, 15, 19, 204-205, 206, 207, 211-212, 213
  regulatory environment, 54, 57, 61-63
  remediation technologies, *viii*, 11, 19, 15, 90, 204-205, 206, 207, 211-212, 213, 220, 222, 223, 225, 228-229, 232, 236, 241, 246
    DNAPLs, 5-6, 9, 137, 146, 150, 182, 188
    metals and radionuclides, 112, 120
  Subsurface Contaminants Focus Area, 12, 204-205, 236, 241, 246
  see also Funding
Court cases, 40
Creosote, 165, 229
Cryogenic barriers, 90, 216, 265, 266, 267

D

Deep-soil mixing, 86, 90, 91, 265-266
Dense nonaqueous-phase liquids, *ix*, 1, 2, 16, 25, 27, 29, 129-201
  air sparging, 7, 135, 147-150, 215, 242
  alcohol flushing, 150-154, 266
  bioremediation, 8, 136, 173-180, 185, 187, 189, 194
  bioventing, 13, 140-141, 246
  cosolvent flushing, 8, 150-154, 194, 241
  Department of Energy programs, 145, 150, 161, 167, 171, 172-173, 183, 228-235, 240, 241-242, 243; *see also* Subsurface Contaminants Focus Area
  dynamic underground stripping, 228-229
  electrical and electrokenetic remediation, 135, 163, 166-168, 171-173, 194, 228, 229, 230-235, 241, 265, 266
  geologic, geochemical, and hydrologic factors, 7, 8, 18, 23, 25-27, 130-135 (passim), 146, 148, 150, 155-159 (passim), 162, 165-166, 177, 185, 188, 192-195
  in situ oxidation, 8, 27, 134, 159-162, 242, 267
  in situ vitrification, 9, 135, 168-171, 194
  joule heating, 166, 169
  Lasagna, 172, 217, 232-235, 248, 265
  perchloroethylene, 29, 52, 53, 145, 156, 161-162, 167, 173-178 (passim), 186, 189-190, 230
  phytoremediation, 9, 180-182
  regulatory environment, 48, 55
  remediation technologies, 4, 7-9, 11, 17, 19, 84-85, 115, 133-135, 140-195, 208, 228-235, 241-242
    cost factors, 5-6, 9, 137, 146, 150, 182, 188
    evaluation, 134-136, 139, 145-146, 149-150, 152-154, 157-159, 161-162, 165-168, 170-173, 177-182, 185-188, 190-195, 243
  soil vapor extraction, 8, 9, 11, 135, 140-146, 150, 167, 168, 178, 193-194, 202, 215, 216, 229, 241-242, 266
    thermally enhanced vapor extraction, 9, 228, 230-231, 243, 265
  steam injection, 8, 163-166, 134, 163-166, 194, 228, 229, 241, 267
  surfactants, 8, 134, 151, 154-159, 194, 241, 266
  trichloroethylene, 29, 48, 52, 53, 106, 149, 156, 157, 161-162, 167, 172-

182 (passim), 186-187, 189-191, 230
see also Chlorinated solvents; Polychlorinated biphenyls; Volatile organic compounds

Department of Defense
remediation technologies, *viii*, 7, 9-11

Department of Energy, *vii, x*, 1-2, 3, 15-37
Accelerated Site Technology Deployment Program, 12, 212, 213-214, 218, 246
contractors and contracting, 9, 10, 15-16, 204, 205, 207, 211, 218-219, 225, 245
cooperative research and development agreements, 10, 225, 227, 236
DNAPLs, 145, 150, 161, 167, 171, 172-173, 183, 228-235, 240, 241-242, 243
extent of cleanup requirements, 2-3, 21-24, 36
sites by state, 253-263
funding, 36-37, 204-205, 207-208, 208-209, 211
Subsurface Contaminants Focus Area (SCFA), 9-11, 12, 13, 19-21, 208, 218-220, 235-236, 240-243 (passim), 244, 247-248
historical perspectives, 1, 2-3, 15, 25, 202, 211
management and managers
general, 11, 13, 16, 205, 207, 217, 244-245
OST, 10, 205-206, 208, 217
metals and radionuclides, 72-128 (passim), 222-228, 240-241, 243, 254-263
Office of Environmental Management (DOE), 36, 207, 211-212, 218
Office of Environmental Restoration, 2-3, 12, 19, 35-36, 212, 244
Office of Science and Technology, *viii-ix*, 2, 10, 20, 203-211, 217-219; *see also* Subsurface Contaminants Focus Area
Office of Technology Development (DOE), 19
remediation technologies, 4, 9-11, 36, 72, 84-85, 101, 202-239; *see also* Subsurface Contaminants Focus Area
regulatory environment, 9, 10, 15-16, 18, 39-71, 206-207, 209, 211, 221, 236
see also specific DOE sites

Dichloroethylene, 20, 48, 174, 176-179 (passim), 186-187, 190, 191

Diethylenetriaminepentaacectic acid, 276

Direct-push technologies, 133, 137

Drinking water, 7
regulatory issues, 3, 39, 42, 49-50, 149-150, 161
maximum contaminant levels, 47-48, 49, 52
Safe Drinking Water Act, 42, 47-48, 52

Dynamic underground stripping, 228-229

E

Ecological risk assessments, 31, 33-35, 67, 73

Economic factors, 203-205
contractors and contracting, 9, 10, 15-16, 204, 205, 207, 211, 218-219, 225, 245
see also Cost and cost-benefit factors; Funding

Electrical and electrokinetic remediation systems, 11
DNAPLs, 135, 163, 166-168, 171-173, 194, 228, 229, 230-235, 241, 265, 266
metals and radionuclides, 6, 8, 78, 81, 112-114, 121

Electrochemical analysis, 82, 95

Energy Policy Act, 257

Enhanced Site Specific Risk Assessment, 65

Environmental Management Science Program, 13, 248

Environmental Protection Agency, 3, 35, 42, 47, 49, 53, 61-62
brownfields, 63, 64
DNAPLs, 51, 145, 152-153, 157-158, 179, 189, 232
management and managers, 49, 62
metals and radionuclides, 102, 113, 115
natural attenuation, 60, 61, 189, 190
technical impracticability waivers, 55-58
SITE, 102, 113, 115, 179

Ethylenediaminetetraacetic acid, 82, 276, 118

Evaluation and evaluation issues, 3, 17, 120-122, 209, 244-245
- Accelerated Site Technology Deployment Program, 12, 212, 213-214, 218, 246
- air sparging, 135, 149-150
- alcohol and cosolvent flushing, 152-154, 194
- bioremediation, 111-112, 121, 177-180, 194
- contaminant characterization, 28-29
- DNAPLs, remediation, 134-136, 139, 145-146, 149-150, 152-154, 157-159, 161-162, 165-168, 170-173, 177-182, 185-188, 190-195, 243
- electrical and electrokinetic processes, 113-114, 121, 135, 167-168, 172-173, 194
- field tests, 8
- in situ oxidation, 134, 161-162, 194
- in situ redox manipulation, 109-110, 121
- in situ vitrification, 9, 99-101, 120-121, 134, 170-171, 194
- metals and radionuclides, remediation, 77-80, 105-107, 109-110, 111-114, 118-121
- natural attenuation, 60, 69, 190-191
- peer review, 12, 67, 68, 209, 244
- permeable reactive barriers, 105-107, 109-110, 121, 183, 185-188, 194
- phytoremediation, 118-120, 121, 181-182
- remediation guidelines, *vii, ix-x,* 13, 16-17, 55, 69, 190, 246
- steam injection, 134, 165-166
- soil flushing and washing, 115-116, 121
- soil vapor extraction, 135, 145-146, 178, 193-194
- solidification and stabilization, 102-103, 121
- subsurface barriers, 94-96, 105-107, 109-110, 120, 121;
- Subsurface Contaminants Focus Area, 12-13, 244-245
- Superfund Innovative Technology Evaluation program, 102, 113, 115, 179
- surfactants, 134, 157-159, 194
- technical impractibility waivers, 57
- *see also* Risk assessment; Sensor technologies

Excavation of contaminated soils, 19, 35

Exposure, *see* Risk assessment

F

Federal Facilities Compliance Act, 40, 41

Federal government, 1, 203-204
- *see also* Funding; Legislation; Regulatory issues; *specific departments and agencies*

Federal Remediation Technologies Roundtable, 13, 246

Fernald Environmental Management Project, 30, 32, 33, 35, 43, 62-63, 73, 213, 259, 266, 267

Foreign countries, *see* International perspectives

Funding, 13
- Accelerated Site Technology Deployment Program, 213-214
- brownfields programs, 64
- Department of Energy, not SCFA, 36-37, 204-205, 207-208, 208-209, 211
- Subsurface Contaminants Focus Area (SCFA), 9-11, 12, 13, 19-21, 208, 218-220, 235-236, 240-243 (passim), 244, 247-248
- *see also* Comprehensive Environmental Response, Compensation, and Liability Act

G

Gasoline, 229, 148, 165

General Accounting Office, 10, 204, 205-206, 207, 208-209, 217, 219

Geologic, geochemical, and hydrologic factors, 7, 16, 21, 27
- DNAPLs, 7, 8, 18, 23, 25-27, 130-135 (passim), 146, 148, 150, 155-159 (passim), 162, 165-166, 177, 185, 188, 192-195
- metals and radionuclides, 25, 27, 76-77, 86, 225, 227
- subsurface barriers, 86, 90, 95-96

GeoVIS, 137

Ground penetrating radar, 138

Guidance for Evaluating the Technical Impracticability of Ground-Water Restorations, 55

Guide to Documenting and Managing Cost

and Performance Information for Remediation Projects, 13, 246

H

Hanford Site, 11, 21, 24, 26, 30, 31, 32, 33, 36, 43, 52, 73, 110, 222-225, 263, 265, 266, 268
Henry's law constant, 130, 142, 145, 147
Historical perspectives
  Department of Energy contractors, 211
  Department of Energy site cleanup, 1, 2-3, 15, 25, 202, 211
  regulatory environment, 3-4, 58, 63
  Superfund, 202
Human health risks, *see* Risk assessment
Hydrogeology, *see* Geologic, geochemical, and hydrologic factors
Hydrosparge VOC sensing system, 137

I

Idaho National Engineering and Environmental Laboratory, 11, 21, 24, 26, 30, 33, 34, 36, 52-53, 213, 214, 222, 225, 243, 257, 265-268 (passim)
Incineration, 18-19
In situ bioremediation, *see* Bioremediation
In situ oxidation, 8, 27, 134, 159-162, 242, 267
In situ redox manipulation, 5, 11, 78, 80, 107-110, 121, 222-225, 243
In situ vitrification, 9, 35, 208-209, 266
  DNAPLs, 9, 135, 168-171, 194
  metals and radionuclides, 5, 78, 80, 96-101, 120-121
Interagency Working Group on Brownfields, 64
International perspectives, 115-116, 161
Internet, 19
Ion exchange systems, 73, 86, 113, 114, 115, 241

J

Joule heating, 166, 169

L

Lasagna, 172, 217, 232-235, 248, 265
Laser-induced fluorescence sensors, 137
Lawrence Livermore National Laboratory, 11, 16, 22, 43, 165, 228, 254, 265-268 (passim)
Legal Environmental Assistance Foundation, 40
Legislation
  Clean Water Act, 42
  Energy Policy Act, 257
  Federal Facilities Compliance Act, 40, 41
  Resource Conservation and Recovery Act, 3, 40-48, 49, 53, 54, 55, 57, 63, 68, 206, 213
  Safe Drinking Water Act (1974), 42, 47-48, 52
  Superfund Act, *see* Comprehensive Environmental Response, Compensation, and Liability Act
  Toxic Substances Control Act, 42
  Uranium Mill Tailings Remediation Control Act, 3, 22, 43, 46-47, 49, 51, 68
  see also Regulatory issues
Light nonaqueous-phase liquids, 137, 157-158, 165
Litigation, *see* Court cases
Local government, *see* State and local governments
Los Alamos National Laboratory, 22, 43, 258

M

Management and managers
  contractors and contracting, 9, 10, 15-16, 204, 205, 207, 211, 218-219, 225, 245
  Department of Energy,
    general, 11, 13, 16, 205, 207, 217, 244-245
    OST, 10, 205-206, 208, 217
  EPA, 49, 62
  regulatory issues, 40, 49, 51, 54-55, 57, 60, 61, 62-63, 65, 69, 245
  site-level, 9, 16, 40, 47, 49, 51, 54-55, 57, 60, 61, 62-63, 65, 69, 204, 212, 225, 244

Maximum contaminant levels, 47-48, 49, 52
Metal pollutants, 1, 16, 29-30, 72-128, 208
  bioremediation, 6, 78, 110-112, 121
  Department of Energy programs, 72-128 (passim), 222-228, 240-241, 243, 254-263; *see also* Subsurface Contaminants Focus Area
  deep-soil mixing, 86, 90, 91, 265-266
  electrical and electrokenetic remediation, 6, 8, 78, 81, 112-114, 121
  geologic conditions, 25, 27, 76-77, 225, 227
  in situ redox manipulation, 5, 11, 78, 80, 107-110, 121, 222-225, 243
  in situ vitrification, 5, 78, 80, 96-101, 120-121
  ion exchange systems, 73, 86, 113, 114, 115, 241
  phytoremediation, 6, 78, 80, 116-120, 121
  pozzolanic agents, 79, 80, 101, 102
  remediation technologies, 4, 5-6, 11, 17, 19, 76, 77, 78-122, 168, 213, 222-228, 241, 243
    cost factors, 112, 120
    evaluation, 77-80, 105-107, 109-110, 111-114, 118-121
  soil flushing and washing, 6, 18, 78, 81, 112, 113, 114-116, 121
  solidification and stabilization techniques, 5, 78, 80, 101-103, 121, 215
  sorption, 76, 78, 80, 86, 103
  speciation, 76, 77, 80
Methanol, 139, 151
Monument Valley, 254
Mound Plant, 43, 73, 213, 214, 259, 265, 267

N

National Contingency Plan, 51
National Priorities List, 43, 44
Native Americans, 31, 32
Natural attenuation, 7, 54, 57, 58-61, 69, 73, 173, 189-191, 215, 243
Natural Resources Defense Council, 40
Naval installation sites, 65, 67-68, 257
Netherlands, 115-116
Neutron probes, 138
Nevada Test Site, 35, 43, 259
Nonaqueous-phase liquids, 120, 137, 142, 157-158, 165
  see also Dense nonaqueous-phase liquids; Light nonaqueous-phase liquids

O

Oak Ridge Reservation, 21, 24, 26, 30, 32-33, 34, 35, 43, 53, 62-63, 73, 214, 261, 265, 267, 268
Office of Environmental Management (DOE), 36, 207, 211-212, 218
Office of Environmental Restoration (DOE), 2-3, 12, 19, 35-36, 212, 244
Office of Science and Technology (DOE), 2, 10, 20, 203-211, 217-219
  see also Subsurface Contaminants Focus Area
Office of Technology Development (DOE), 19
Oil, *see* Petroleum
Oxidation, *see* In situ oxidation

P

Paducah Gaseous Diffusion Plant, 43, 172, 234-235, 257, 265
Pantex Plant, 43, 262, 268
Peer review, 12, 67, 68, 209, 244
Penetrometers, 11, 83, 133, 216, 222, 225-228, 243
Perchloroethylene, 29, 52, 53, 145, 156, 161-162, 167, 173-178 (passim), 186, 189-190, 230
Permeable reactive barriers, 5, 78, 80, 81, 103-107, 121, 136, 182-188, 194, 213, 215, 217, 242, 265
  see also In situ redox manipulation
Petroleum, 65, 144, 166, 175-176, 217
Phytoremediation
  DNAPLs, 9, 180-182
  metals and radionuclides, 6, 78, 80, 116-120, 121
Plutonium, 25, 28, 31, 52, 76, 77, 111, 256
Polychlorinated biphenyls, 29, 42, 47, 102
Polycyclic aromatic hydrocarbons, 137
Potassium permanganate, 159-162, 183-185, 187
Pozzolanic agents, 79, 80, 101, 102

Pressurized injection, barriers, 88-89
Privately owned sites, 40, 203
Public exposure, *see* Risk assessment
Public involvement, 2, 19, 67, 68, 206
Pump-and-treat systems, 3, 7, 18, 27, 36, 85, 104, 150, 212, 215, 222, 225, 228-229
Push-in technologies, 133, 137

R

Radar, *see* Ground penetrating radar
Radio-frequency heating, 167-167, 230
Radionuclides, 1, 16, 27-28, 29-35, 72-128, 208, 222
  deep-soil mixing, 86, 90, 91, 265-266
  Department of Energy programs, 72-128 (passim), 222-228, 240-241, 243, 254-263; *see also* Subsurface Contaminants Focus Area
  electrical and electrokenetic remediation, 6, 8, 78, 81, 112-114, 121
  geologic conditions, 25, 27, 76-77, 86
  in situ redox manipulation, 5, 11, 78, 80, 107-110, 121, 222-225, 243
  in situ vitrification, 5, 78, 80, 96-101, 120-121
  ion exchange systems, 73, 86, 113, 114, 115, 241
  phytoremediation, 6, 78, 80, 116-120, 121
  pozzolanic agents, 79, 80, 101, 102
  remediation technologies, 4, 5-6, 11, 17, 19, 74, 76, 77, 78-122, 171, 213, 214, 222-228, 241, 243
    cost factors, 112, 120
    evaluation, 77-80, 105-107, 109-110, 111-114, 118-121
    see also specific technologies infra and supra
  soil flushing and washing, 6, 18, 78, 81, 112, 113, 114-116, 121
  solidification and stabilization techniques, 5, 78, 80, 101-103, 121, 215
  sorption, 76, 78, 80, 86, 103
  speciation, metals and radionuclides, 76, 77, 80
  standards, 48, 50, 52-53; *see also* Uranium Mill Tailings Remediation Control Act
  see also Plutonium; Uranium
Regulatory issues, 2, 3-4, 39-71, 240, 245
  alternate concentration limits, 48-49
  applicable or relevant and appropriate requirement, 48, 49, 51
  Department of Energy, 9, 10, 15-16, 18, 39-71, 206-207, 209, 211, 221, 236
  drinking water, 3, 39, 42, 49-50, 149-150, 161
    maximum contaminant levels, 47-48, 49, 52
    Safe Drinking Water Act, 42, 47-48, 52
  historical perspectives, 3-4, 58, 63
  management and managers, 40, 49, 51, 54-55, 57, 60, 61, 62-63, 65, 69, 245
  maximum contaminant levels, 47-48, 49, 52
  new technologies and, 9, 10, 15-16, 206-207, 209, 211, 221, 236
  pump-and-treat remediation, 18
  records of decision, 46, 52, 53, 54, 56, 61-63, 69
  secrecy requirements, 25
  soil screening level, 49-50
  Subsurface Contaminants Focus Area, 9, 10, 12, 39, 54, 55, 57, 61, 63, 68, 69, 221, 236, 243, 245
  technical impracticability waivers, 54-58, 69
  see also Legislation
Redox manipulation, *see* In situ redox manipulation
Remediation technologies, 1-2, 16-17, 55, 56-70
  conventional, limitations of, 17-19
  cost factors, 11, 19, 15, 90, 204-205, 206, 207, 211-212, 213, 220, 222, 223, 225, 228-229, 232, 236, 241, 246
    DNAPLs, 5-6, 9, 137, 146, 150, 182, 188
    metals and radionuclides, 112, 120
  DNAPLs, 4, 7-9, 11, 17, 19, 84-85, 115, 133-135, 140-195, 208, 228-235, 241-242
    cost factors, 5-6, 9, 137, 146, 150, 182, 188
    evaluation, 134-136, 139, 145-146, 149-150, 152-154, 157-159, 161-

162, 165-168, 170-173, 177-182, 185-188, 190-195, 243
see also specific technologies infra and supra
Department of Defense, *viii*, 7, 9-11
Department of Energy, *viii-ix*, 4, 9-11, 36, 72, 84-85, 101, 202-239; *see also* Subsurface Contaminants Focus Area
metals and radionuclides, 4, 5-6, 11, 17, 19, 76, 77, 78-122, 168, 213, 222-228, 241, 243
cost factors, 112, 120
evaluation, 77-80, 105-107, 109-110, 111-114, 118-121
see also specific technologies
site-level management and managers, 9, 16, 40, 47, 49, 51, 54-55, 57, 60, 61, 62-63, 65, 69, 204, 212, 225, 244
technical impracticability waivers, 54-58, 69
see also Evaluation and evaluation issues; Regulatory environment; Sensor technologies; Technical assistance; *specific technologies*
Remediation Technologies Development Forum, 13, 248
Resource Conservation and Recovery Act, 3, 40-48, 49, 53, 54, 68, 206, 213
brownfields, 63
technical impracticability waivers, 55, 57
Risk assessment, general, *viii*, 3, 13, 17, 21, 29, 30-35, 54, 65-68, 69, 241
brownfields, 64
ecological, 31, 33-35, 67, 73
excavation of soils, 18
metals and radionuclides, 74
new technology, 9, 10
subsurface barriers, 9, 10, 11
Subsurface Contaminants Focus Area, 9, 10, 11
wildlife, 33-35
Risk-based corrective action, 65-68, 69
Risk-based screening level, 4
Rocky Flats Environmental Technology Site, 21, 24, 26, 30, 32, 34, 35, 43, 53, 73, 214, 256

S

Safe Drinking Water Act, 42, 47-48, 52
Sandia National Laboratories, 43, 60, 73, 213, 230, 254, 259, 267, 268
Savannah River Site, 19, 21, 24, 26, 30, 32, 34, 35, 43, 53, 73, 150, 161, 214, 261, 265, 266, 267
Seismic techniques, 95, 138
Semivolatile organic compounds, 8, 99, 168, 193-194, 230
Sensor technologies, 83-84, 94, 95, 133, 137-140, 213, 216, 221, 225, 227, 228, 242, 266, 267
see also specific technologies
Sequential extraction procedures, 82-83
Single-extraction procedures, 83
Site characterization and analysis penetrometer system, 222, 225-228, 243
Site Screening and Technical Guidance for Monitored Natural Atttenuation at DOE Sites, 60
Site technology coordination groups, 207
Site Technology Deployment Program, 211-212
Soil flushing and washing, 6, 18, 78, 81, 112, 113, 114-116, 121
Soil screening level, 49-50
Soil vapor extraction, 8, 9, 11, 135, 140-146, 150, 167, 168, 178, 193-194, 202, 215, 216, 229, 241-242, 266
thermally enhanced vapor extraction, 9, 228, 230-231, 243, 265
Solidification and stabilization techniques, 5, 78, 80, 101-103, 121, 215
Solvents
for DNAPL remediation, 134
see also Chlorinated solvents; Cosolvent flushing
Sorption, 76, 78, 80, 86, 103
Southern States Energy Board, 209, 211
Speciation, metals and radionuclides, 76, 77, 80
Spectroscopy, 77, 83, 137
State and local governments, 39, 40, 140, 206, 240
brownfields, 63-64
monitored natural attenuation, 60-61
risk-based corrective action, 65
Standards, *see* Regulatory issues
Steam injection, 8, 163-166, 134, 163-166,

194, 228, 229, 241, 267
Strontium, 28, 29, 48, 102, 113
Subsurface barriers, 5, 9, 11, 12, 36, 84-96, 120, 136, 188, 213, 222, 225, 242-243
  cryogenic barriers, 90, 216, 265, 266, 267
  geologic, geochemical, and hydrologic conditions, 86, 90, 95-96
  permeable reactive barriers, 5, 78, 80, 81, 103-107, 121, 136, 182-188, 194, 213, 215, 217, 242, 265; *see also* In situ redox manipulation
  risk assessments, general, 9, 10, 11
  trenching, 5, 43, 86-88, 106, 115, 121, 143, 152, 177, 183, 188
Subsurface Contaminants Focus Area, 2, 7, 9-13, 19, 202-236, 240-248
  cost effectiveness, 12, 204-205, 236, 241, 246
  funding, 9-11, 12, 13, 19-21, 208, 218-220, 235-236, 240-243 (passim), 244, 247-248
  regulatory environment, 9, 10, 12, 39, 54, 55, 57, 61, 63, 68, 69, 221, 236, 243, 245
Superfund, *see* Comprehensive Environmental Response, Compensation, and Liability Act
Superfund Innovative Technology Evaluation program, 102, 113, 115, 179
Surfactants, 8, 134, 151, 154-159, 194, 241, 266

T

Technetium, 5, 28, 29
Technical assistance, Subsurface Contaminants Focus Area, 10, 12, 13, 246-247
Technical impracticability waivers, 54-58, 69
Technology Practices Manual for Surfactants and Cosolvents, 151
Trenching, 5, 43, 86-88, 106, 115, 121, 143, 152, 177, 183, 188
Tetrachloroethylene, 48
Thermal desorption volatile organic compound sampler, 137
Thermally enhanced vapor extraction, 9, 228, 230-231, 243, 242, 243, 265
Toxicity characteristic leaching procedure
Toxic Substances Control Act, 42
Trichloroethane, 181
Trichloroethylene, 29, 48, 52, 53, 106, 149, 156, 157, 161-162, 167, 172-182 (passim), 186-187, 189-191, 230
Tritium, 28, 29, 208

U

Uranium, 25, 28, 29, 32, 33, 48, 53, 73, 74, 76, 77, 111, 213, 254-263
Uranium Mill Tailings Remediation Control Act, 3, 22, 43, 46-47, 49, 51, 68

V

Vadose zone, 84, 94, 95, 130, 131, 132
Video imaging system, (GeoVIS), 137
Vinyl chloride, 29, 48, 173, 176, 177, 179, 186-187, 190
Vitrification, *see* In situ vitrification
Volatile organic compounds, 137, 140-146, 150, 161, 178, 186, 213
  see also Semivolative organic compounds

W

WASTECH, 13, 246
Weather, *see* Climatic conditions
Wildlife, 33-35
World Wide Web, *see* Internet